The Ultimate
Nutrient
Glutamine

Judy Shabert, MD
Nancy Ehrlich

Avery Publishing Group

Garden City Park, New York

This book has been written and published strictly for informational purposes, and in no way should it be used as a substitute for recommendations from your own medical doctor or health-care professional. All the facts in this book came from medical files, clinical journals, scientific publications, personal interviews, and the personal-practice experiences of the authorities quoted or sources cited. You should not consider educational material herein to be the practice of medicine or to replace consultation with a physician or other medical practitioner. The authors and publisher are providing you with the information so that you can have the knowledge and can choose, at your own risk, to act on that knowledge.

Cover designer: William Gonzalez
In-house editors: Barbara Conner and Elaine Will Sparber
Typesetter: Bonnie Freid
Printer: Paragon Press, Honesdale, PA

Library of Congress Cataloging-in-Publication Data

Shabert, Judy.
   The ultimate nutrient, glutamine : the essential nonessential
amino acid / Judy Shabert, Nancy Ehrlich.
      p. cm.
   Includes bibliographical references and index.
   ISBN 0-89529-588-1
   1. Glutamine—Physiological effect. 2. Glutamine—Health aspects.
I. Ehrlich, Nancy. II. Title.
QP562.G55S52 1994
612'.01575—dc20   93-45656
                                                       CIP

10 9 8 7

Contents

To Doug.

Acknowledgments

Many people supported and encouraged our endeavors to create this book, but two individuals in particular stand out. Doug Wilmore considered the idea of a book on glutamine to be a sound and worthwhile idea. His encouragement and critical comments throughout the project enabled us to write the most current and accurate book on glutamine possible. Janet Lacey spent hours of her time reading, typing, checking references, and giving supportive comments about the book over the last two years. We believe that without her help, the work on this book could not have progressed as smoothly as it did.

Kristann Wilmore was available from the inception of the project to give psychological support about the worthiness of our ideas.

Several people took time from their very busy schedules to discuss with us their clinical experiences and their ongoing research on glutamine—Dr. Wiley W. Souba, Dr. Suzanne Klimberg, Dr. Paul Schloerb, and Dr. John Rombeau. For readers interested in a scientific overview of the physiology and biochemistry of glutamine, we recommend Dr. Souba's book *Glutamine: Physiology, Biochemistry and Nutrition in Critical Illness* (Austin, TX: R.G. Landes Company, 1993).

We also wish to thank Dr. Stephen F. Lowry, who provided helpful information on cytokines.

We appreciate the information on the manufacture of amino acids given to us by the Ajinomoto Company through Dr. Eiji Goto and Arlene Peters.

We would like to thank our friends and relatives who read drafts of the book and lent critical comments, including Raphael and Frieda Ehrlich, Betty Shabert, Ruth Rahn, Cheryl Jacobs Ehrlich, Bob Lyle, Adam Levy, Toby Bilanow, Stephen McNab, and Lori Mogol. Henry Rahn and Dr. Arthur Adrian spent hours of their time analyzing the manuscript for grammatical and technical accuracy; to them, we are extremely grateful.

Our publisher, Rudy Shur, deserves a special thank-you for the enthusiasm he has shown since he first read parts of the manuscript. It has been a pleasure working with him and with the staff at Avery Publishing Group.

Judy Shabert, M.D., R.D.
Nancy Ehrlich

Foreword

What is glutamine, and why is it important? First, glutamine is a nutrient found in the food we eat. More specifically, it is an amino acid, which serves as a necessary building block for the protein in the body. Protein makes up the important structures of the body; it is the major component of the muscles, heart, liver, bowel, kidneys, and other organs. In addition, protein controls the function of the body; for example, all of our enzymes and most of our hormones, which control the body's chemical reactions, are proteins. Unfortunately, when we become sick, all these proteins become quite vulnerable and may break down; we become weak, our resistance falls, we become susceptible to infection, and the body is slow to repair itself. Physicians try to rebuild or maintain body protein in sick individuals, but this has been a difficult goal. It is as if something were missing from the food provided.

Glutamine has been like a long-lost relative to the investigators working to learn how to rebuild body protein. These scientists generally knew that glutamine was there, but they did not know much about it—especially about what it does. It has been in only the past ten years or so that we have found that glutamine is:

- The most common amino acid in the body.
- A key to the metabolism and maintenance of muscle.
- The primary fuel, or energy source, for the entire immune system.
- Essential for deoxyribonucleic acid (DNA) synthesis, cell division, and cell growth, which are all necessary for wound healing and tissue repair.
- The primary nutrient for the cells that line the gastrointestinal tract.
- Important for neutralizing toxins in the body.

When we are well, the body converts the food we eat into enough glutamine to support many of these functions. However, when we are sick, the body cannot keep up with all these demands for glutamine. More important, there is so little glutamine in food that barely enough is available even when we are healthy. And glutamine is rarely present in the specialized food provided to hospital patients.

This book tells you about the important discoveries relating to glutamine and helps you to apply this information to your individual situation. It allows you to incorporate this important nutrient in your recovery from illness, in the maintenance of your body, and in the prevention of disease.

The costs and risks of glutamine are quite low; the benefits to specific groups of people are quite substantial. In spite of this, you should discuss the information in this book with your physician should you choose to incorporate glutamine in your plan to care for your body.

The growing interest and research in this field mean that more discoveries and information about glutamine are forthcoming. Thus, this book should serve as only an introduction to what we will ultimately know about the exciting amino acid glutamine—which is essential for your health.

Douglas W. Wilmore, M.D.
Harvard Medical School

Preface

A major discovery was made more than two hundred years ago. Lemons and limes were found to prevent scurvy in sailors who ate them during long voyages. Now, at the threshold of the twenty-first century, scientists have discovered the value of the amino acid glutamine. The supplemental use of glutamine may be as important to people recovering from illness as lemons and limes were to sailors hoping to avoid scurvy.

Much of the current information on glutamine, however, is scattered in a wide variety of scientific journals. Furthermore, the numerous studies on glutamine are not readily available to professionals and are certainly not common knowledge to the public. Because glutamine is proving to be so significant in the treatment of so many illnesses, we wanted to correct this situation and write a comprehensive yet easily understandable book for individuals both with and without a scientific background.

Most chapters of *The Ultimate Nutrient, Glutamine* include an anecdote related to us by a physician who has used glutamine in the care of patients. The first chapter is an overview of glutamine and the research on it. The second chapter discusses the role of muscle in providing the body with glutamine. The third chapter describes the significance of glutamine to the gastrointestinal tract. Chapter Four reveals the unique role the liver plays in using and supplying the body with glutamine, as well as the important relationship of glutamine to glutathione, the most significant free-radical scavenger in the body. The fifth chapter deals with the immune system and how glutamine has been shown to be helpful to individuals with compromised immune systems, such as patients with acquired immunodeficiency syndrome (AIDS). Chapter Six explores the role of glutamine as a mood enhancer, and Chapter Seven looks into the tremendous potential of glutamine in the care of cancer patients. Chapter Eight discusses aging, vitamins, and glutamine. And Chapter Nine explains how to incorporate glutamine into your life under the supervision of a health-care provider. The book ends with a glossary and a comprehensive list of references.

We hope this book will stimulate further interest in glutamine, the nonessential amino acid that many scientists are now beginning to hail as essential.

1

Glutamine– Essential for Health

Healthy people need specific nutrients in their food in order to stay healthy. People who are sick or stressed have different nutrient requirements. One nutrient that must be added to the food of sick or stressed people is glutamine, which scientists have only recently realized is important. Now, the debility associated with many illnesses—whether mild or serious—can be dramatically reduced.

In this age of high technology, it seems incredible that one nutrient can be causing such a stir in the scientific and medical communities. Researchers all across the United States and Europe, and in such distant places as Japan, are realizing the importance of a nutrient called glutamine.[1-3]

Recent discoveries surrounding glutamine, an amino acid, have linked it to the most important functions of the body's vital organs and systems. Among these life-sustaining roles, glutamine helps the body make more muscle. Inadequate supplies of glutamine are responsible for the wasting of muscle and for the weakness that accompanies a fever or other stressful illness.[4,5] Scientists believe that glutamine is the major energy source of the intestines, allowing the normal functioning of the gastrointestinal tract. The nutrient also helps control fluid loss from the intestines.[6,7] Glutamine is also believed to be the second most important fuel for the cells lining the colon and to help clear the body of waste through the kidney and liver.[8] However, perhaps glutamine's most important function concerns the immune system. Glutamine is involved with the multiplication of selected white cells, which strengthen the body's defense system, and it also helps other immune cells to kill bacteria.[9,10] Glutamine aids in healing wounds. It supports pancreatic growth. It maintains and supports glutathione, an important antioxidant. As an ingredient for forming glutathione, glutamine may help prevent the replication of the virus that causes acquired immunodeficiency syndrome (AIDS).[11]

How can all of these important functions be performed by a single amino acid? Why have we been unaware of the importance of glutamine for such a long time? This book will take you through the history of this special nutrient, helping you to understand what it is capable of doing. It will also show you how you can integrate glutamine into your diet when the need arises.

THE IMPORTANCE OF NUTRIENTS IN THE BODY

When we consume food, we take in major nutrients, vitamins, and minerals, all of which contribute to the health and maintenance of the body. Foods are composed primarily of three major nutrient groups—protein, carbohydrate, and fat.

Proteins, made up of amino acids, function as the building blocks of muscle and other structural components of the body. They also make up all of the body's enzymes and many important hormones (known as peptides). Proteins also provide energy and repair damaged tissue. When we ingest protein in food, however, our bodies must first break down the protein into its amino acid components and then reassemble the components into new protein. In the human body, twenty important amino acids are used to make proteins. From this group of twenty, nine amino acids are considered *essential* and eleven are considered *nonessential*. By *essential*, we mean that the body cannot create the amino acid on its own; the amino acid must be provided by the food we eat. By *nonessential*, we mean that the amino acid can be manufactured in sufficient quantities by the body's own tissues. The essential amino acids are histidine, isoleucine, leucine, lysine, methionine, phenylalanine, threonine, tryptophan, and valine. The nonessential amino acids include alanine, arginine, aspartic acid, asparagine, glutamic acid, glycine, proline, and serine. Cystine and tyrosine are nonessential amino acids that may substitute for essential amino acids under appropriate conditions.

Glutamine was considered a nonessential amino acid for many years. But recent research has changed this view, and since the 1980s, glutamine has been considered a *conditionally essential* amino acid.[12] This means that under normal circumstances, the body can make (synthesize) adequate quantities of the amino acid, but in times of stress such as fever, illness, dieting, or chemotherapy, the body cannot make as much as it requires. An additional amount of the amino acid must be taken to prevent a deficiency.

The carbohydrate foods include the sugars and starches, and in their best form, they are very rich in fiber. Carbohydrates primarily generate the energy we need to use our muscles and brains. In general, however, when carbohydrates are ingested, the energy they provide is used by the body rapidly. For example, when an individual chops wood or runs a marathon, carbohydrate energy is what is quickly utilized. After carbohydrate energy is depleted, fat becomes the energy source of choice. Excess energy from ingested carbohydrate and protein food is stored in the body as fat (adipose tissue). The storage form of carbohydrate is glycogen. Fats also serve as the structural foundation of many hormones in the body, such as cortisol, the stress hormone, and estrogen, the feminizing hormone.

All of the food components just described—carbohydrates, fats, and amino acids—can be broken down further to the basic elements of carbon, oxygen, and hydrogen. Amino acids can also be broken down to nitrogen and some other elements such as sulphur. These elements are connected by chemical bonds, which can be thought of as links in a chain. It takes energy to form these bonds; for example, when the body needs to manufacture new glucose (sugar), it must utilize and properly link together the basic molecules of carbon, hydrogen, and oxygen to form the larger glucose molecule. When a bond (link) is broken, energy is released. This energy is what the body uses to perform all of its functions, from such microscopic tasks as digesting food to such obvious ones as moving the extremities. It is through this process of breaking down compounds and building up compounds that we maintain our bodies. In scientific terms, this process is known as *metabolism*. When the body's tissue breaks down at a rate greater than that at which it is built up, the body is in a state known as *catabolism*. The opposite—putting on muscle and growing—is known as *anabolism*.

WHAT MAKES GLUTAMINE UNIQUE

Amino acids, which are derived from proteins, are different from carbohydrates and fats because, in addition to containing carbon, hydrogen, and oxygen, they also contain a nitrogen atom, which is in a highly specific structure. Glutamine, moreover, contains two nitrogen atoms. This extra nitrogen may be the factor most responsible for the uniqueness of glutamine. In scientific jargon, glutamine is known as a nitrogen shuttle, a substance that picks up and drops off nitrogen around the body. Through this process of shuttling nitrogen from tissue to tissue, glutamine works at various sites on such tasks as clearing poisonous wastes like ammonia; adding a nitrogen atom to make deoxyribonucleic acid (DNA), the genetic material of life; and building muscle.

It took researchers a very long time to consider the importance of glutamine, in part because it was chemically so difficult to analyze. In addition, glutamine was believed to be nonessential, that is, it was believed that the body always synthesized enough glutamine on its own. With our current level of understanding, it seems rather naive not to have studied glutamine more closely. It is the most abundant amino acid in the body, including in the brain, skeletal muscle, and blood. In fact, it is proving to be the most significant amino acid known.

The researchers who first realized glutamine's unique importance were not medical doctors but scientists who were trying to grow cells in the laboratory. During the 1950s, Harry Eagle tried to get human and animal cells to grow by adding glucose to the cells' nutrient solutions.[13] He did this because so many cells of the body depend on glucose as a source of energy. He discovered, however, that glucose was not enough. After trying many different substances, he found that glutamine actively supported the growth (proliferation) of immune and other cells in cell culture dishes. Not only did the nutrient solution containing glutamine support cell growth, but if glutamine was not added to the cell culture media, there was no proliferation of cells at all.

Little was done with glutamine until the 1970s, when investigators working at the National Institutes of Health (NIH) found that glutamine, not glucose, is the most important nutrient for the intestinal tract.[14] In the early 1980s, physicians discovered that during the stress associated with an illness, the body breaks down its own muscle and produces large quantities of glutamine. The amino acid is then carried by the bloodstream to sites where it is most needed—the kidney, the intestines, the liver, and the immune cells—to aid in the specialized functions of combating illness and repairing the body. All of these functions are performed at the expense of breaking down muscle. It is no wonder that weakness and loss of muscle mass so commonly occur following illness.

In the early 1980s, a group of scientists at Harvard University set out to determine the exact components of muscle breakdown during stressful conditions.[15] They placed catheters (small tubes) in arteries and veins in the legs of laboratory animals. Arteries are blood vessels that carry blood to tissues. Veins are blood vessels that carry blood away from tissues (in this case, the veins drained the animals' leg muscles). By analyzing the concentration of amino acids going to the muscles via the arteries and coming from the muscles via the veins, the investigators could determine the exact amino acid components that were being liberated by the leg skeletal muscles. First, they took measurements in animals that were awake and resting. Then, they administered a synthetic stress hormone to simulate the stressful condition that animals and humans experience at times of illness or surgery. According to the results, when the stress hormone was administered, there were rapid increases in the breakdown of muscle tissue and in the release of amino acids from the muscles. The quantity of amino acids released from the legs increased approximately threefold.

The new discovery showed that up to one-third of the amino acids released at times of stress is in the form of glutamine. Further studies have revealed that glutamine is synthesized by muscles as they break down (atrophy) in times of stress.[16]

Similar findings were observed in the same animal model following general anesthesia and a very simple abdominal operation. Attempts were then made to reverse this catabolic state. When standard intravenous amino acid solutions, like the solutions currently used in hospitalized patients, were infused, muscle breakdown continued and was only minimally attenuated. When glutamine alone was infused, however, there was less muscle loss. When glutamine was infused with a full complement of other amino acids, muscle breakdown was essentially prevented.[17]

Studies in humans have shown very similar findings. Patients who have major surgery or severe injury or burns have a rapid breakdown of skeletal muscle and release of amino acids.[18] Glutamine constitutes a major component of the products of this muscle breakdown and is produced in three ways. A small portion is produced simply by the release of glutamine that had been floating freely within muscle cells. An even smaller portion comes from the breakdown of muscle protein into its constituent amino acids, of which glutamine represents only a small fraction (4 percent). The major portion is produced when some of the other amino acids in the muscle cells re-form into glutamine molecules.

European investigators have examined the effect of

Many years ago, the woman had had cancer of the cervix. Radiation therapy for treatment of the cancer had severely damaged her intestines. Now, the woman had to have most of her intestines removed in a massive operation. She was left with only a few feet of damaged small intestine connected to the last few feet of her colon. Most people who undergo this type of operation are severely debilitated after the surgery. Most require years of intravenous feeding, since their intestines no longer function normally. If such patients eat solid food, the meal is immediately followed with severe watery diarrhea. These people often remain too weak to return to work. This patient's surgeon, however, had heard about glutamine's positive effects on the gastrointestinal tract and on the immune system, and he had instructed her to start taking glutamine before the operation and to continue taking it afterward. To everyone's surprise and pleasure, the woman, although on a restricted diet, was soon able to eat solid foods again and after a few months was able to return to work.

adding glutamine to the nutrient solutions infused into postsurgery patients.[19] For example, when individuals undergo an operation such as the removal of the gallbladder (a procedure known as cholecystectomy) or more complex procedures such as the removal of a cancerous portion of the colon, there is an increased breakdown of skeletal muscle, a release of amino acids into the bloodstream, and a fall in the skeletal muscle levels of glutamine. If standard intravenous nutrient solutions are infused, there is very little impact on these processes. When glutamine is added to the amino acid solution, however, the muscle breakdown rates are greatly diminished. When the rates of skeletal muscle protein synthesis were measured, it was found that muscle synthesis fell dramatically after an operation; however, this could be restored by the inclusion of glutamine in the intravenous solution. One can imagine how much stronger the patients must have felt when they were not weakened by major muscle loss.

Only recently have researchers been giving glutamine to critically ill patients. It has been found that muscle breakdown can be prevented with glutamine administration and that the protective functions of the other organs during periods of stress are actually enhanced.

After reading this book, you may realize that the amino acid glutamine might be helpful to you or to someone you know. It is a unique, special, and important nutrient, and must not be confused with other substances that have similar-sounding names (see "What Glutamine Is *Not*," below). Glutamine will almost certainly prove to be the most important nutrient of the twenty-first century. Its cost is reasonably low, it can be obtained without a physician's prescription, and it can be easily incorporated into the diet.

What Glutamine Is *Not*

Glutamine is not glutamic acid.
Glutamine is not glutamate.
Glutamine is not monosodium glutamate (MSG).
Glutamine is not gluten.
Glutamine is not glutathione.

2

Preventing
Muscle Breakdown

The obese middle-aged woman agreed to participate in the glutamine study. She was told that after her gallbladder surgery, she would be given either standard intravenous nutrition or intravenous nutrition supplemented with glutamine—but she would not know which she was getting. She was told that if she was given glutamine, her body would not have to use as much protein for healing as it would have to without glutamine; she would be "sparing her body protein." She was told that recovery from a gallbladder operation usually takes about six weeks. Much to the woman's amazement, she felt strong, vigorous, and ready to return to work only one week after her operation. She felt without a doubt that she had received glutamine, and she now understood what "protein sparing" meant.

We live in an era of health consciousness. The muscle strength of someone like bodybuilder Arnold Schwarzenegger or the muscle endurance of someone like runner Jackie Joyner is highly prized in our society. Our grandparents did not focus on being strong, although many people of their era were strong because they performed physical labor in farming or other strenuous jobs. Many of our parents turned to the cities for occupations in which a strong body was not necessary. Gradually, our society became physically unfit. Then, in the 1970s and 1980s, running, aerobics, weight lifting, and healthier diets came into vogue. These changes have contributed in part to the increased longevity we are seeing in our nation's people.

All forms of exercise require muscle strength. The amount of muscle we have and its capacity for endurance determine to a large extent our level of excellence in many areas of physical prowess. As we age, our muscle mass gradually decreases, as does our strength. The amount of strength we have can have a great impact on our ability to continue independent living. Let us try to understand what muscle is and what it does, and what role the amino acid glutamine plays in maintaining a healthy, active body.

MUSCLE—A BRIEF REVIEW

In the body, there are three different types of muscles— smooth, cardiac, and striated. Smooth muscle surrounds tubular structures such as the blood vessels and intestines. It is wisplike, thin, and simple in structure. An example of smooth muscle function is the movement of the intestines to propel food through the digestive system after a meal. These muscles are not under our conscious control but are controlled automatically by a part of the nervous system called the autonomic nervous system.

Cardiac muscle is found in the heart and is under autonomic control, although it responds to a variety of local factors, such as amount of blood in the heart chambers and amount of acid buildup in the blood. Cardiac muscle is generally thicker than smooth muscle, and its structure is somewhat more complex.

The third type of muscle, striated muscle, is the focus of this chapter. More is known about the role of glutamine in striated muscle than in either of the other two muscle types. Striated muscle is found on the skeleton; it connects one joint to another. (The biceps muscle, for example, runs from the shoulder joint to just below the elbow.) Skeletal muscle is specialized to allow rapid contractile movement of the body. Normally, it is in a state of relaxation. It can be thought of as a window shade, which is in the relaxed position when it is down and becomes contracted when force is put on it and it rolls up. Unlike either smooth or cardiac muscle, skeletal muscle is under voluntary control. Another difference is that skeletal muscle represents approximately 15 to 40 percent of an individual's body weight, whereas smooth or cardiac muscle represents less than 2 percent of body weight at the most. The amount of muscle mass in a given individual is determined by genetics, gender, use (exercise), and state of health.

THE ROLE OF GLUTAMINE IN MUSCLE

Muscle cells are fused together into myofilaments. Within each cell are the contractile fibers actin and myosin. When muscle is stimulated by a nerve to contract, these fibers interact and shorten, resulting in the contraction of the muscle. The actin and myosin are surrounded by cytoplasm, which is the storage center for many of the cell's structures. Lying between the actin-myosin bundles within the cytoplasm are the disc-shaped mitochondria. Their function is to make energy for the cell, much like a furnace makes energy for heating a house.

In muscle cells, extra large amounts of glutamine are found floating freely.[1] By floating freely, we mean that it is not linked to any other amino acid, so it is readily available to the body in times of need. In fact, glutamine is the most common free amino acid in the body, accounting for approximately 60 percent of all the free amino acids. (The glutamine in the cells is known as the intracellular pool of glutamine.) Furthermore, glutamine is found in extremely high concentrations in the bloodstream (plasma).[2] Although glutamine is considered to be at a high level in the plasma, its concentration there is still only one-thirtieth of the concentration of the intracellular glutamine pool. This high concentration of glutamine in the muscle cells allows the cells to discharge large quantities of glutamine into the blood when glutamine is needed at other organ sites.

During the 1970s, as scientists were trying to understand the relationship of amino acids and protein to skeletal muscle, they found that the amount of an amino acid in the blood does not correlate very well with the amount of the amino acid in the cytoplasm of the skeletal muscle (the intracellular pool).[3] It is thus difficult to know if glutamine deficiency is present because it cannot be determined just by measuring the level of the amino acid in the plasma. This makes it necessary to perform the rather uncomfortable and invasive procedure of muscle biopsy (taking a sample of muscle tissue) to diagnose glutamine deficiency.

When glutamine concentrations are measured simultaneously in muscle and blood, the concentrations can be quite low in the muscle but not in the blood. However, if blood levels of glutamine are low, it is a sure indication that the individual is glutamine deficient.[4,5]

When an individual is healthy, the food that he or she eats can supply the body with the proteins needed to maintain body structure and function. As discussed in

Chapter 1, protein is composed of subunits known as amino acids. Basically, when food is digested, its protein is broken down to amino acids, which are then absorbed. These amino acids are further metabolized in the liver to be used as a body fuel like glucose, reassembled into a new body protein, or transformed into other compounds including other amino acids. Because of glutamine's unique structure, it is often utilized for all three functions—a fuel, a new body protein, and other important compounds and amino acids.

SICKNESS AND GLUTAMINE

When individuals are metabolically stressed, they become catabolic, which means that their tissue is breaking down. People can become catabolic in a variety of ways, for example, from flu, dieting, starvation, infection, injury, or burns. Using steroids like prednisone or engaging in strenuous physical activity also produces metabolic stress. When an individual is metabolically stressed, the muscle produces significantly more glutamine in order to maintain blood levels. Nevertheless, concentrations of glutamine within the muscle cells may fall by at least 50 percent.[3] If enough protein is not taken in through food to meet the body's demands, the muscle begins to break down to supply the body with glutamine. This glutamine is then transported through the bloodstream to promote the healing of wounds,[6] to aid cells in fighting infections,[7,8] or to support the gastrointestinal tract.[9-11] This is the reason for the muscle wasting we experience during times of illness or stress.

To prove that muscle wasting is secondary to glutamine deficiency, scientists in the 1980s conducted studies in which they observed what happens in muscle tissue and blood and to amino acid levels when the body is stressed.[12] To simulate stress, they gave animals a stress hormone similar to the adrenal stress hormone cortisol, since the body produces cortisol naturally in most stressed states. The scientists found that when the animals were stressed, there was a threefold increase in the release of glutamine from their muscles.

In their next study, the scientists measured the levels of amino acids in patients' muscle cells before and three days after major surgery, which is a known metabolic stress.[5] The study revealed that the blood level of glutamine remained stable, even though the level of glutamine in the muscle cells fell by 50 percent.

Once it was determined that intracellular glutamine is reduced when an individual undergoes such stress as major surgery, scientists wanted to know what happens to glutamine levels during more significant stress, such as major trauma followed by massive infection (sepsis).[3] The scientists did muscle biopsies on patients who had suffered multiple trauma injuries, again when they developed sepsis, and again in the recovery period. Not surprisingly, they found that intracellular glutamine levels were low after the injury and in the presence of sepsis. Furthermore, during the convalescent period, when other amino acid levels were returning to normal, muscle glutamine levels remained low, indicating that there is a slow recovery of glutamine levels back to normal.

Investigators have additionally shown that other stressors, such as burns, deplete intracellular glutamine to low levels.[6] They have also found that during prolonged starvation, blood levels of glutamine cannot be maintained, and eventually, the blood also becomes low in glutamine.[13]

The researchers who conducted these studies felt that

glutamine's role in the body was shown to be fairly straightforward. The blood levels of glutamine remain stable at the expense of the stores of glutamine in the muscle cells. This is in fact what happens, but further studies have revealed that there are other unique properties related to glutamine and muscle.

Researchers used clever methods to determine what was specifically going on with glutamine in skeletal muscle.[14,15] They attached one catheter to an artery, which is a vessel that carries blood into a tissue such as a muscle, and another to a vein, which is a vessel that carries blood away from a tissue. This was done in the forearm of the subject, so the blood flow to and from the muscle could be observed. The researchers discovered that glutamine was at a lower level in the blood that approached the muscle than it was in the blood that left the muscle. This proved two things. First, muscle is a storage depot for glutamine, making glutamine available to the body during periods of stress; and second, muscle actually synthesizes glutamine for the rest of the body to use.

HOSPITAL PATIENTS AND GLUTAMINE

The enzyme glutamine synthetase, which is contained in high levels in muscle, becomes activated when the body needs more glutamine. Commonly, this enzyme is activated when a hospitalized person has doctor's orders to receive nothing by mouth. Usually, the patient is not allowed to eat and is given fluids intravenously, or through a vein. The fluid infusion is known as a standard intravenous (IV). This fluid usually contains a small amount of glucose and some sodium chloride (salt). The glucose is given because the brain and muscles use it as their primary fuel. This solution has minimal effects against muscle breakdown.

It has long been recognized that there is a rapid atrophy of muscle tissue during illness or hospitalization. Physicians have attempted to reverse the process by feeding patients early in the course of illness or by giving them a special intravenous solution containing far greater energy sources (carbohydrate and fat) and protein (amino acids) than included in a standard IV. The special solution is infused through a vein in the arm, neck, or chest and is known as total parenteral nutrition (TPN).[16] Unfortunately, even when patients received as much as twice their usual requirements in calories and protein, muscle breakdown could not be prevented.[17]

The implications from this are very significant. Our present methods of nutritional support for hospitalized patients do not prevent the breakdown of muscle following stress. However, if stressed or injured people could preserve their muscle mass, the road to recovery would be much shorter and the return to good health assured.

The most common experiment to test whether the provision of glutamine to individuals undergoing a metabolic stress can prevent or attenuate muscle loss has been to give glutamine intravenously. This was done in studies in the United States at Harvard University, in Germany, and in Sweden. Following surgery or bone-marrow transplantation, patients were given either the usual TPN solution, which does not contain glutamine, or a similar TPN solution containing glutamine. Those patients who received glutamine had a significantly smaller decrease in their intracellular glutamine concentrations, had improved synthesis of their skeletal muscle protein, and had less loss of total body protein than did the individuals receiving the usual solution.[18-20]

This effect can be explained by the relationship between intracellular glutamine concentrations and the ability of the muscle to make protein. When the level of glutamine in a skeletal muscle decreases, the ability of the muscle to make protein (that is, more muscle) also falls and the amino acids, which are the building blocks of protein, rapidly leave the cell. When the concentrations of glutamine return to normal, the muscle's ability to make more muscle also increases and muscle loss is reduced.[21-23] If a person takes supplemental glutamine during a stressed state, the ability of his or her muscles to make protein is supported, and the breakdown and wasting of skeletal muscle mass are prevented.

ATHLETES, BODYBUILDERS, AND GLUTAMINE

Athletes and bodybuilders have a special interest in glutamine. From data on treating patients, it is obvious that glutamine plays a significant role in protecting muscle mass. One reason that athletes or strenuous exercisers may lose muscle mass or fail to gain it is acidosis, which occurs with strenuous exercise. Exercise causes metabolic reactions to take place and a surplus of acid to be formed in the body. After exercise, the body compensates for the acid and gets rid of the acidic byproducts through the action of glutamine on the kidneys. Glutamine formed from muscle breakdown goes to the kidneys and donates a molecule to neutralize the positive charge from the acid. A recent study in animals revealed that supplemental glutamine can prevent acidosis and reduce muscle breakdown.[24]

To date, there are no well-documented studies on the effect that supplemental glutamine has in muscle building, since we are just at the early stages of understanding its significance in illness. We do know without a doubt, however, that supplemental glutamine prevents muscle breakdown. It may not be long before studies are done to show conclusively that glutamine also enhances muscle development in healthy individuals.

Glutamine is not contained in powdered protein supplements for bodybuilding. The labels on protein supplements list glutamate (Glu), not glutamine (Gln). There is a big difference. Glutamine contains the extra nitrogen that gives it its special properties; glutamate does not. Glutamine is not in these products because heat is used during the processing of powdered supplements, and glutamine is destroyed at high temperatures. Thus, even if a product starts out with glutamine, by the time it goes through the processing to make it into a powder, the glutamine has been destroyed.

Most liquid protein supplements also do not contain additional glutamine. At room temperature, glutamine gradually degrades and loses its potency. The actual quantity of glutamine cannot be ensured unless the formula is kept refrigerated. However, glutamine powder can be added to a liquid protein supplement, which should then be refrigerated and used within several days.

The studies performed using seriously ill or postoperative patients have important implications for nonhospitalized patients who are in metabolically stressed states, whether from illness, heavy exercise, infection, dieting, or another cause. Taking as little as 15 grams (15,000 milligrams, or 3¾ teaspoons) to 20 grams (20,000 milligrams, or 5 teaspoons) of glutamine may maintain muscle stores of glutamine and prevent net muscle breakdown. Since glutamine represents only a small fraction of the protein we take in as part of our regular diets (4 to 5 percent), and since much of the glutamine present in foods is inactivated by exposure to heat or cooking, a supplement of glutamine may be necessary to meet the requirements of metabolic stress.

3

Healing the Stomach and Intestines

The baby had had colic and bloody diarrhea since she was five months old. Crohn's disease, the doctors had said, making her the youngest patient diagnosed with this ailment in the country. Steroids had been prescribed, and they helped, but the baby did not grow. By the time the child was six years old, her now-desperate mother took matters into her own hands. She had heard from other patients about the importance of glutamine for bowel growth, so at night, she gave her child a liquid feeding containing glutamine. During the day, she added glutamine powder to the child's cold drinks. After five months of glutamine, the child's bowel symptoms reversed, and the physician reduced the steroid medication. Unlike the five previous times when the anti-inflammatory drugs had been reduced, the child's symptoms did not recur. The steroids were stopped, but the glutamine was continued. The child has had only minor gastrointestinal symptoms and has been off steroids for one and a half years. Additionally, she has had a significant growth spurt.

In the last chapter, we spoke of the muscle breakdown (catabolism) that occurs during times of stress, so that the rest of the body can be supplied with glutamine. The first significant discovery about glutamine was that it exists in a high concentration in muscle cells and that the muscle enzyme glutamine synthetase actively makes glutamine in an effort to maintain high blood concentrations. During stress, in spite of glutamine synthesis rising twofold to fourfold, glutamine concentration decreases by 50 percent in muscle and by 30 percent in blood.

Where is the glutamine going? If muscle is supplying glutamine, what tissues are consuming it?

THE DIGESTIVE SYSTEM—A BRIEF REVIEW

In this chapter, we will focus on the digestive system, an area of the body in which glutamine is utilized to a very large extent. At the core of the digestive system is the gastrointestinal tract. The long tube of the gastrointestinal tract begins at the mouth, where food enters the digestive system, and ends at the anus, where waste products are expelled. In between the mouth and the anus are the esophagus, the stomach, and the small and large intestines. Supportive organs that work with the gastrointestinal tract include the liver, the pancreas, and the gallbladder. Collectively, these organs are known as the digestive system. Glutamine plays a major role in the health and well-being of these structures.

Digestion—the breaking down of food into smaller parts (ultimately, microscopic particles)—begins in the mouth. Although it is obvious that our teeth crush and divide food, it is not so obvious that enzymes are produced and released in the mouth to break down food at a molecular level. If you imagine your favorite food, you will probably notice an increase in the secretions in your mouth. These secretions, known as saliva, contain the enzymes that facilitate the process of food breakdown.

From the mouth, food is moved by strong propulsion down the esophagus and into the stomach, where further digestion takes place in the stomach's highly acidic environment. Hydrochloric acid, along with the mixing action of the stomach, rapidly turns food into a liquid mass. At the base of the stomach is a valve, the pyloric sphincter, which opens to allow food to travel into the duodenum, the first segment of the approximately twenty-foot-long small intestine. Much activity occurs in the duodenum—the breakdown of protein into amino acids, the conversion of starches into simpler carbohydrates, and the breakdown of fats.

The luxuriant lining of the small intestine is known as the mucosa. The mucosa is composed of fingerlike projections called villi, which are where the actual intestinal cells, the enterocytes, are located. The enterocytes are some of the most rapidly multiplying cells in the body. The presence of digesting food inside the tube (lumen) of the intestinal tract stimulates the growth of these cells. (This stimulus is known as luminal drive.)

The enterocytes absorb digested food and transport the nutrients obtained from it to tiny blood vessels inside the villi that lead to the circulatory system via the liver. The energy that allows the enterocytes to absorb nutrients and otherwise function normally comes to the cells via the bloodstream; the primary nutrient for these cells is glutamine. The enterocytes contain an enzyme, glutaminase, the sole function of which is to divide glutamine into two other compounds—glutamate, which is an amino acid, and ammonia. The glutamate is broken down further in order to obtain five molecules of adenosine triphosphate (ATP), the energy-packed phosphate compound necessary for most of the cellular reactions that occur in the body.

In addition to the small intestine, the gallbladder, pancreas, and liver support the process of digestion. The gallbladder is a saclike structure about as large as a small coin purse. It stores a substance called bile acid, which is produced by the liver. Bile acid acts like soap or detergent, breaking down large globules of fat into smaller and smaller globules until they are small enough to be absorbed into the bloodstream through the wall of the small intestine. This process of forming tiny globules is known as emulsification.

The pancreas, which sits behind the stomach and looks like a long, pink, soft tongue with a pointed tail, functions as both an endocrine and an exocrine organ. An *endocrine organ* produces a hormone or substance that travels through the bloodstream and acts on a site in the body far away from the organ. An *exocrine organ* secretes a substance locally that acts on nearby sites. The pancreas is an endocrine organ because it produces the hormone insulin, which flows from the pancreas into the bloodstream to be utilized around the body to facilitate the entry of glucose into cells. The pancreas also has exocrine functions because it produces alkaline secretions and enzymes that are discharged into the intestines to help digest food. Glutamine is known to serve as an important fuel source for the exocrine functions of the pancreas and possibly for the endocrine functions as well.[1]

The liver, which will be discussed in greater detail in Chapter 4, is the recipient of the nutrients released into the bloodstream. The liver is positioned on the right side of the body under the right lung. A very large and vital organ, it has the amazing capacity to regenerate itself if part of it is injured or removed. In fact, it is now possible for transplant surgeons to remove part of one person's liver and transplant it into the body of another individual. However, if an individual's liver suffers significant damage or the cells of the inner structure are destroyed, the person will die because the liver is needed for so many important functions. These functions include the processing and storing of simple sugars, fats, and amino acids, and the reassembling of these substances as the body needs them. For example, the liver takes glucose, turns it into glycogen, a more complex sugar, and stores it. It then turns the glycogen back into glucose when the body's need for energy increases.

As mentioned on page 23, the breakdown of glutamine in the enterocytes of the small intestine yields glutamate and ammonia. As we all know, ammonia is poisonous to the human body, and here we can begin to see another important role of the liver—it detoxifies or neutralizes many substances, including ammonia, which would be poisonous if present in the body in large quantities. The intestine has a direct link to the liver through the portal vein, so poisonous substances go from the intestines to the liver. In the liver, the poisons are extracted from the blood before they have a chance to enter the circulatory system, where they would be carried to the body's tissues, including the brain.

The lower part of the gastrointestinal tract—the large intestine and rectum—functions mainly to extract water from the mass of nondigestible food and recycle it back into the blood. In the lower intestine, important bacteria live in abundance. These bacteria produce vitamin K, which is necessary for the clotting mechanisms of the blood.

Unlike certain other systems of the body, such as the

nervous system and the cardiac system, the digestive system is rarely given its due. It is not seen as mysterious, like the brain, or considered to be vital, like the heart. If it is thought of at all, it is viewed with disdain as the repository of our self-indulgences (such as a second helping of dessert) or as the site where our stresses become manifest in the form of gastritis or ulcers. Even when the digestive system is given respect, it is usually thought of as the site where nutrients are sent off to the bloodstream to be utilized by other, "more important" organs. Not until the last fifteen years, during which time the role of glutamine in the digestive system has become better known, have scientists begun to look at this system with increasing admiration.

THE ROLE OF GLUTAMINE IN THE INTESTINES

It appears that the intestines play a central role in the body during illness or catabolism, and glutamine metabolism is basic to the intestines' functions.[2] The first important discovery concerning this was made by Herbert Windmueller, a pharmacologist at the National Institutes of Health.[3] Dr. Windmueller wanted to study the absorption characteristics of antibiotics and other drugs in the small intestine. To do so, he had to keep a small segment of animal intestine alive by perfusing it with a solution containing the appropriate nutrients. There was only one problem. The usual solutions for maintaining perfused organs contain primarily glucose, and these solutions could not keep the small piece of intestine alive.

Dr. Windmueller reasoned that the perfusion solution did not contain the appropriate nutrients, and he set out to discover what the intestine "eats"—or metabolizes—to stay alive. Windmueller and his associates found that glutamine was the major fuel burned by the enterocytes. Glucose sugar, which we commonly eat and which is provided to patients in the hospital, was virtually not utilized as a fuel source by the intestine. These findings have been confirmed in a variety of studies utilizing both animals and humans.[4-6]

INTESTINAL REPAIR AND GLUTAMINE

Glutamine's role in the intestines was discovered only in the past fifteen years, but we should now realize how the discovery relates to the care of sick patients. Until recently, the intestines were thought to be inactive during illness, and most health-care practitioners believed they should not be utilized during illness or surgery. It had been thought that the intestines needed to "rest" to repair themselves. However, the main nutrient necessary for intestinal repair—glutamine—is not included in many feeding formulas and is absent in all present intravenous feeding solutions. In addition, when an individual does not eat, what really happens is that the cells lining the intestines, the mucosal cells, atrophy because they lack the stimulation they usually receive from digesting food. Because the stimulus from digesting food is not present and glutamine is lacking, the usual luxuriant, thick lining of the intestines gives way to thin, denuded tissue, which is easily eroded, ulcerated, or permeated by bacteria.[7] Thus, by withholding food during illness in an attempt to allow the gut to repair itself and by providing only a glucose-based solution for support, the opposite of what is wanted occurs—rather than being repaired, the intestine is injured. The lining of the intestine becomes thin, and bacteria from the intestines penetrate the intestinal wall and enter adjacent tissues and eventually, in some cases, the bloodstream.

ILLNESS AND GLUTAMINE

When an individual is in a normal state of health, the mucosal cells of the intestines, situated on the villi, rapidly grow and regenerate themselves, utilizing glutamine as their primary energy source. During illness, the demand for energy by the intestines increases dramatically. We know now that these tremendous energy demands of the intestines during stress are met at the expense of the body's own protein.[2] Muscle breaks down into amino acids, and through the action of glutamine synthetase, glutamine is formed. The glutamine travels via the circulatory system to the intestines, where it is taken up to support the continuous growth (turnover) of the mucosal lining.

Additional glutamine may be used by the immunological cells that are closely associated with the intestines. (See Chapter 5 for a discussion of the role of glutamine in the immune system.) The mucosal lining of the gut, which plays an important role in maintaining health, is both a physical barrier (because the cells of the villi do not allow bacteria to pass through to the bloodstream) and an immunological barrier. If bacteria or viruses do manage to get past the physical barrier, special white blood cells deep in the villi engulf and destroy them. However, when the normal supply of glutamine to the intestines is impaired, such as when a person does not eat, both components of the mucosal barrier falter. The barrier becomes permeable, or "leaky," allowing organisms and toxins to go through the intestinal wall. This passage through the gut wall is known as bacterial translocation.

In addition, the immunological cells do not capture the microorganisms as effectively. The bacteria are then picked up by the circulatory system and transported throughout the body. This is felt to be one cause of the hypermetabolism seen in severe infection or injury.

Usually, when an individual undergoes a severe stress, such as a major burn, a predictable response occurs. Initially, there is a stress response, which means that certain hormones—namely, the glucocorticoids, which are glucagon and adrenaline—are discharged into the bloodstream. Then, the individual becomes hypermetabolic, that is, his or her metabolism starts running at a high rate and the body's tissue breaks down rapidly. This reaction may be secondary to a leaky intestinal barrier, which allows bacteria and bacterial poisons (endotoxins) into the bloodstream. Scientists wondered if they could blunt this response by beginning to feed patients sooner than normal after a major stress, thereby better preserving the intestinal barrier. They studied laboratory animals, then humans, to prove that the intestinal barrier could be preserved.

The researchers studied injured guinea pigs.[8] They divided the guinea pigs into two groups. One group was not fed for three days after injury; these animals became hypermetabolic, and their gut mucosa atrophied. The other group was fed soon after the injury; these animals became only slightly hypermetabolic, and their gut mucosa was maintained. The information gained from this work has changed the way burn patients are cared for in the hospital. Burn patients are now given food by mouth, or intravenously if necessary, within hours of their admission to the hospital. This puts less demand on their muscles to supply glutamine, and the patients do not become as wasted or weak.

There are occasions when people are unable to eat for extremely long periods, either because of an illness, such as inflammatory bowel disease, or a therapy, such as chemotherapy or radiation for cancer. Under these conditions, the capacity of the intestine to maintain itself starts to wane. The levels of glutamine in the blood begin to decrease because the demand exceeds the supply, cell replication becomes much slower, muscle is significantly wasted, and the intestines severely atrophy. It appears that the negative impact of illness is far greater and occurs far earlier in the intestines than in any other organ in the body. The recovery time is significantly prolonged as well.

CHEMOTHERAPY, RADIATION, AND GLUTAMINE

Scientists, having realized that glutamine plays such a critical role in intestinal function, wondered if they could somehow diminish the stress reaction and thereby spare the body's muscle from wasting. Studies were done to prove that glutamine supplementation intravenously does indeed diminish the loss of the mucosa of the intestines.

In one study, experimental animals were not allowed to eat but instead were given all their nutrients in a total parenteral nutrition (TPN) solution devoid of glutamine, the same solution used in hospitals today.[9] As expected, the animals' mucosal linings showed significant atrophy. When glutamine was added to the solution, the amount of atrophy decreased by half. Another study involved chemotherapy, which is used in the treatment of cancer.[10] Basically, chemotherapy works against cancerous growth by destroying rapidly growing cells. However, chemotherapy attacks and destroys not only cancer cells but also cells of the intestinal tract, which are the fastest growing cells in the body. This accounts for much of the nausea, vomiting, and diarrhea suffered by people who are being treated for cancer.

Scientists wanted to see if they could protect the gastrointestinal tract from the destruction caused by chemotherapy medication.[11–13] They gave experimental animals extremely high doses of 5-fluorouracil (5-FU) or methotrexate (MTX), two chemotherapy drugs. Some of the animals also received glutamine. When the animals were autopsied, those who had received the glutamine had a significantly healthier intestinal lining, as evidenced by taller villi, than did the animals that did not receive glutamine. The glutamine protected the animals from the toxic side effects of the drugs. Not only did the intestines of the animals that received glutamine do better, but many of the animals that were not given glutamine died. There were similar findings when animals were given radiation in high doses similar to those often given to patients with cancer in the abdominal cavity.[14] Again, the experimental animals treated with glutamine fared much better.

What about humans? Obviously, scientists cannot autopsy human patients to see if glutamine therapy was effective on the intestinal tract, but some information can be gained indirectly. As previously mentioned, chemotherapy is very toxic to the gastrointestinal tract, as is radiation. Why not look at people who have had high doses of both chemotherapy and radiation? There is currently such a group in the United States—patients undergoing bone-marrow transplantation. These patients are first given extremely high doses of chemotherapy, then their entire bodies are irradiated. They are kept in isolated, germ-free rooms while the transplanted bone marrow regenerates. In spite of this precaution, however, many of the patients become infected. The microbes are thought to come primarily from an injured intestine. Therefore, some positive effects could be expected from glutamine, and indeed, some very positive outcomes were noted.

The study was conducted by a group at Harvard University under the direction of Dr. Douglas Wilmore.[15] The patients were divided into two groups—those receiving standard TPN, which does not contain glutamine, and those receiving TPN supplemented with glutamine. The glutamine group received 40 grams (40,000 milligrams, or 10 teaspoons) of glutamine per day in small doses over twenty-four hours. The patients receiving glutamine had a significantly decreased incidence of infection and were released from the hospital earlier than were the patients not receiving glutamine. These gains may not seem so extraordinary, but for people isolated in an intensive care unit and under constant observation by hospital staff, this represents a huge benefit. In this case, the hospital bills of the patients receiving glutamine were reduced by an average of $21,000. (For a more detailed discussion of this study, see page 63.)

STOMACH ULCERS, DIARRHEA, AND GLUTAMINE

So far, we have discussed the role of glutamine in maintaining the integrity of the small intestine, but glutamine serves other important functions in the gastrointestinal tract as well. Japanese scientists have discovered that glutamine is an effective antiulcer drug for the stomach.[16] First, they created ulcers in one group of rats by giving them aspirin. Then, they totally prevented the development of these drug-induced ulcers in another group of rats by giving them glutamine. In other studies, they discovered they could enhance the healing of peptic (stomach) ulcers by giving test subjects oral glutamine. Glutamine is now the most popular antiulcer drug in Asia.

Glutamine has also been found to be important in diminishing the loss of electrolytes and water from the intestines during diarrhea. Rotavirus, which belongs to the group of viruses causing the most infant diarrhea worldwide, was given to experimental animals.[17] All the animals developed diarrhea, but the animals given glutamine fared better; they showed better absorption of water than did the animals not given glutamine because the handling of sodium chloride by the intestinal mucosa was enhanced. Glutamine is now being studied as a possible therapy for infants and children who have diarrhea. Oral glutamine could help enhance water and salt uptake into the body and could help lessen diarrhea.

Finally, glutamine also has been found to be an important nutrient for the large bowel (the colon) and can provide fuel to maintain the normal function of the mucosal lining cells of the colon.[18]

Now, given these various effects of glutamine on the gastrointestinal tract, how can we use this amino acid to resolve gastrointestinal problems?

One group with upper gastrointestinal problems is people who must take nonsteroidal anti-inflammatory drugs (NSAIDs) for such conditions as arthritis, painful menstrual periods, and severe headaches. The names of some of the NSAIDs are ibuprofen (such as Motrin and Advil), Ponstel, Anaprox, and Naprosyn. Because Japanese scientists have shown that glutamine can prevent drug-induced gastritis (inflammation of the stomach), glutamine administration may be a logical choice for people who are bothered by ulcer formation in the stomach or the gastritis usually seen after taking an NSAID or other drug. Currently, some patients are taking 1,000 milligrams (1 gram, or ¼ teaspoon) of glutamine one-half hour before taking the NSAID. The positive responses have been heartening.

INFLAMMATORY BOWEL DISEASE, SHORT BOWEL SYDROME, AND GLUTAMINE

Another group of people with extreme gastrointestinal problems who may benefit from glutamine is patients with inflammatory bowel disease (IBD), which causes breakdown in the intestinal mucosa, inflammation, and infection. Both British and Canadian investigators gave IBD patients liquid diets containing small amounts of glutamine.[19] At the end of two weeks, most of the patients no longer had diarrheal stools and their abdominal pain had disappeared. Ultimately, the patients were able to return to normal food. Other IBD patients were treated with altered diets and glutamine powder. Their symptoms, including diarrhea, also improved.

A group of investigators in Japan devised an experiment in which the inflammatory bowel disease usually seen in humans was simulated in animals.[20] Some of the animals consumed diets that contained glutamine and some of the animals consumed glutamine-free diets. All of the animals developed ulcers in the small and large intestines, but the animals that were given 27 percent of their protein dose in the form of glutamine had only one-third as many ulcers as did the animals that had no glutamine supplementation. Furthermore, the mucosal height, which is a reflection of the health of the intestine, was 20- to 25-percent higher in the glutamine-supplemented animals.

A group of investigators in Ireland conducted a similar experiment.[21] They gave animals regular laboratory chow or one of two nutrient solutions—one with glutamine or one without glutamine. Their results showed that the animals fed the regular chow had less weight loss than did the animals fed the nutrient solution. However, the animals given the nutrient solution with glutamine had the fewest ulcers in their colons.

It certainly appears that glutamine had an effect in preventing the formation of ulcers in these experimental models of inflammatory bowel disease. It is not unreasonable to speculate that glutamine might also prevent the formation of ulcers in humans with inflammatory bowel disease.

Patients who have had most of their small intestines removed for a medical reason such as severe bowel injury, colitis, or Crohn's disease develop a condition known as short bowel syndrome. Often, an individual is left with just inches of small intestine, which under normal conditions is twenty-one feet long. Traditionally, doctors have treated patients who have a short bowel with total parenteral nutrition. The patients must receive all fluids and nutrients by vein. They often still eat food, but it frequently passes through their bodies without being absorbed and they experience massive diarrhea, often consisting of up to three quarts of fluid per day. One can imagine how limited these people are in participating in normal daily activities, being in constant fear of not having toilet facilities available.

At the present time, a group of health-care providers in Boston is treating such patients with glutamine (first by vein and then orally), with a special high-fiber, low-fat diet, and in some cases, with growth hormone.[22] Patients who have been on intravenous feeding for six to eleven years have been miraculously making a transition to eating a modified diet. They have also been having two or three semisolid bowel movements per day, instead of numerous episodes of watery diarrhea. Many of these patients feel that control of their lives is being given back to them. Currently, this therapy is in its infancy, and most gastroenterologists (doctors who care for people with gastrointestinal diseases) are skeptical until they see, over the long term, how the patients improve and how they are able to live fuller lives.

A recent study at the Mayo Clinic in Minnesota impresses us once again with the importance of glutamine for the intestines and the immune system.[23] Patients with ulcerative colitis often have their colons removed. A little pouch is made from a segment of small intestine, and this artificial colon is connected to the rectum. Individuals with this condition have six to eight bowel movements a day, and about one-third of them develop "pouchitis," which is infection or inflammation of the pouch where the fecal matter is stored. In the Mayo study, patients who had recurring pouchitis that did not respond to the usual antibiotic treatment were randomized into two groups. One group put a small amount of glutamine (1,000 milligrams, which is equivalent to 1 gram or ¼ teaspoon) in their pouches twice a day. The other half placed another substance in their pouches. Not surprisingly, the glutamine group showed a much greater improvement of the inflammation than did the other group. The patients using glutamine were able to eat without incurring infection or requiring long-term antibiotic therapy.

Finally, there is another group of people who have as many gastrointestinal problems as do the groups so far

The patient, who needed to get his food in liquid form because he had a gastrointestinal problem, had developed a heart condition. He had cardiac arrhythmias (irregular heartbeats), and he needed a medication called quinidine. But quinidine can cause diarrhea, and in this patient, the consequences were disastrous. The doctors tried switching the patient's liquid diets, but the diarrhea got worse and worse. Then, the doctors gave the patient a liquid feeding that contained glutamine. Within thirty-six hours, the diarrhea decreased by two-thirds. The patient continued the glutamine-supplemented feedings, and he was able to take the heart medication and to resume his normal life.

discussed, but for a different reason. These are individuals infected with the AIDS virus, which directly infects the cells of the gastrointestinal mucosa. AIDS patients also suffer secondary infections from other bacteria and viruses. The symptoms include massive diarrhea, with the loss of many quarts of water per day. In many cases, the patients become so weak and disabled that they are unable to leave home. In some clinics, such individuals are being given 40 grams (40,000 milligrams, or 10 teaspoons) of glutamine per day in small doses over twenty-four hours. One such patient, whose only change in therapy was the addition of glutamine, reported that he had complete resolution of his diarrhea and was once again able to leave the confines of his home.[24]

Obviously, glutamine is absolutely essential for the digestive tract. When glutamine levels are low, people are at risk for such conditions as sepsis and hypermetabolism. Glutamine can aid in the healing of stomach or intestinal ulcers, can enhance water absorption during diarrhea, and can reduce infection in patients at risk for immunological stress.[25]

4

Supporting
the Liver

The young man was known to be depressed, but no one thought he would try to commit suicide. At about four o'clock one afternoon, he swallowed most of the pills in his medicine cabinet, including a large bottle of Tylenol and some phenobarbital capsules. A neighbor found him collapsed and barely breathing at eight o'clock that evening, and he was rushed to the hospital. Every appropriate method of treatment and life support was initiated—the young man was placed on a ventilator, his stomach was pumped, his intestines were purged, and he was given intravenous fluids. Two hours later, his blood studies showed that toxic and predictably lethal levels of Tylenol were present. He could not accept the oral antidote for Tylenol overdose because his intestines were not working, but without such treatment, his liver would not function and he would die. Then, a staff doctor recalled hearing of animal studies that used a new amino acid solution containing glutamine to reverse Tylenol toxicity. The glutamine so-

lution was infused intravenously, the patient was gradually weaned from the ventilator, and over the next several days, his liver function tests were monitored. Several days later, no abnormalities in the patient's liver functions had yet developed. The glutamine solution had saved his life.

Next to the brain, the liver may be the most "thinking" organ in the body. Unlike the heart, which basically has one function—to contract and expand repeatedly throughout its lifetime—the liver has a multitude of functions. Some scientists estimate that the liver is responsible for nearly five hundred different processes. It is a very large organ. In male adults, it weighs approximately three and a half pounds. It sits under the right lung and stretches across almost the width of the body.

THE LIVER—A BRIEF REVIEW

Nearly all nutrients enter the bloodstream from the intestines. The first path nutrients take as they travel away from the intestines is the portal vein, which goes into the liver. Nutrients cannot circulate around the body until they have been processed in the liver. There, nutrients are extracted from the blood, repackaged, and either stored or sent into the general circulation. The liver also receives blood that is enriched with oxygen and comes from the general circulation through the hepatic artery. The liver is the only organ that receives its blood supply from two sources. The blood from these two sources—the portal vein and the hepatic artery—mixes in blood vessels within the liver called sinusoids. Here, white blood cells and special immune cells found only in the liver and called Kupffer's cells inactivate or engulf any foreign material, such as bacteria, that might inadvertently have entered the blood through the intestines. The sinusoids also have perforations that allow nutrients to be extracted by liver cells, or hepatocytes. Hepatocytes function differently, depending on where they are located along the sinusoids. The periportal hepatocytes, which are located around the portal vein, have different functions than do those around the hepatic vein, which takes blood away from the liver.

An array of processes occurs in the hepatocytes. The basic one is the breakdown of sugars into their simplest form, glucose. Unused glucose is then repackaged as glycogen, the storage form of sugar, and is stashed away in the hepatocytes until the body needs it. When the body is in a fasting state, such as when a meal is skipped or during a night's sleep, the stored glycogen is reconverted into glucose and sent out of the liver to supply various organs with the energy they need to function appropriately. The brain is extremely dependent on glucose as its energy source, and muscle also relies on it. The liver's ability to convert glycogen into glucose is especially important during periods of fasting, such as at night when we are asleep, or when our energy needs are great, such as during strenuous exercise.

In another important process, the hepatocytes receive amino acids from the intestines and rebuild them into proteins. In fact, the liver can convert fats and sugars into some amino acids, and vice versa, as long as the element nitrogen is available to make the amino acid. The hepatocytes can also convert nitrogen to urea, a waste product that is then eliminated in the urine.

THE ROLE OF GLUTAMINE IN THE LIVER

The liver is like a magician, converting different food compounds from one to another. It should therefore be no surprise that the liver plays an important role in the metabolism of glutamine. As we learned in previous chapters, muscle produces glutamine through the action of the enzyme glutamine synthetase, and the intestine consumes glutamine through the action of the enzyme glutaminase. The liver, however, is both a glutamine producer and a glutamine consumer.[1] It would be of no great benefit to the body if the enzymes for production and for consumption of glutamine were closely related in a given cell and were equally active. The net gain of glutamine would end up being cancelled out by the net loss. So, the enzymes glutamine synthetase and glutaminase are distributed within the liver in such a way that when the body needs glutamine, glutamine can be produced, and if the body does not need glutamine, glutamine can be converted to other amino acids or broken down to nitrogen, which is then converted to urea and excreted as waste.

Whenever an amino acid is broken down, an atom of nitrogen is released. Although nitrogen is an essential element in all amino acids and is vital to the body, an excess of nitrogen can be harmful. It can form ammonia, which is toxic to brain tissue. The liver rids the body of excess nitrogen in the form of ammonia in one of two ways—it uses the nitrogen to form urea, which is transported in the blood and ultimately excreted as urine, or it attaches the nitrogen to the amino acid glutamate and thereby forms glutamine. The process begins when blood containing ammonia enters liver sinusoids and interacts with urea-making enzymes in the hepatocytes near the portal vein. The reaction between the ammonia and the enzymes produces large amounts of urea, but some ammonia escapes and travels through the rest of the liver sinusoids. As it travels in the sinusoids, it eventually comes in contact with the glutamine-making enzyme glutamine synthetase. Thus, not only does the system eliminate ammonia, but it also forms glutamine. Glutamine is unique among the amino acids because it contains not just one but two nitrogen atoms. This makes it especially helpful in the body as a nitrogen shuttle, picking up nitrogen and transporting it to another part of the body to be utilized in another function. Glutamine is essential to the body for recycling nitrogen.

GLUTATHIONE, FREE RADICALS, AND GLUTAMINE

The liver detoxifies other harmful compounds, too. For the detoxification process, it uses many important metabolic mechanisms but relies especially on glutathione. Glutathione is manufactured in the liver from three amino acids—glutamate, cysteine, and glycine.[2,3] The glutamate portion of the glutathione compound is derived in large part from glutamine. The liver contains the most abundant supply of glutathione in the body. Glutathione is exported in the bloodstream and also stored in tissues such as the red cells, the intestinal tract, the immune cells, and the lungs.

To better understand glutathione, we should digress for a minute and discuss free radicals, which can injure our organs, and antioxidants, which can protect us from such injury.[4–8] Beta-carotene is an antioxidant that many people have heard about because it is frequently discussed in newspapers and popular magazines. However, some people may not fully understand how antioxidants work and why they are important. When foreign particles invade the body, white blood cells seek them out and discharge powerful chemicals to destroy them. The white blood cells use oxygen to form these powerful chemicals, but unfortunately, because the oxygen contained in the molecular structure of these chemicals is unstable, some harmful substances are formed as well. These substances are known as free radicals, radical oxidants, or oxygen free radicals because they are the byproducts of chemical reactions that involve oxygen. The best-known of the free radicals are the hydroxyl and peroxide molecules, which are similar to the lye that is used in household cleaning solutions or the hydrogen peroxide that is used for bleaching and disinfecting. One can well imagine how destructive these substances can be to the tissues of the body. Free radicals are known to injure cell membranes and to cause defects in DNA, the genetic material of a cell. In addition, free radicals are thought to contribute to such illnesses as arthritis, atherosclerosis (hardening of the arteries), cancer, and cataracts, as well as to the aging process.

Free radicals are constantly being produced in the body, but we are protected from their damage by *anti*oxidant enzyme systems and scavenger compounds. In addition to its other functions, glutathione is the main scavenger of toxic oxidants in the body. For many years, researchers studying glutathione metabolism have overlooked the role glutamine plays in the formation of glutathione. Because glutathione is so critical to the antioxidant system, scientists have been trying to figure out how the body can be induced to make more of it. One of the three amino acids in glutathione is cysteine, which can serve when needed as an essential amino acid. Scientists have given large quantities of cysteine to test subjects to promote the formation of glutathione. Indeed, this has worked quite well, as the following example demonstrates, but it is not the complete answer.

Some people attempt suicide with overdoses of the pain reliever acetaminophen (a common brand is Tylenol). Initially, the overdose does not seem to cause much harm. After two to three days, however, destruction of the liver becomes apparent. This occurs because the acetaminophen binds tightly with glutathione and renders it inactive. With no antioxidants present in the liver, the additional free radicals that are generated attack the liver cells and cause cell injury and cell death. Because the liver has so many essential functions, this destruction means certain death of the patient unless glutathione can be formed to inactivate the damage done by the acetaminophen. Fortunately, there is a way to do this. A medicine known as Mucomyst contains a form of cysteine. The patient drinks the Mucomyst, which increases the amount of glutathione formed and frequently, if given early enough, saves the life of the patient.[9] Unfortunately, no active form of Mucomyst is available in the United States for intravenous use.

What happens if, as with the young man at the beginning of this chapter, the person's digestive tract is malfunctioning and there is poor absorption of the Mucomyst? Unfortunately, until very recently, the person was doomed to die, possibly being saved only by an emergency liver transplantation. Now, however, a team of investigators at Harvard University has established an important relationship among glutamine, glutathione, and liver injury. In the initial study, prior to the testing of acetaminophen, experimental animals were given near-lethal doses of the antitumor drug 5-fluorouracil (5-FU).[10] One-half of the animals also were given an intravenous feeding solution containing

Warning for People With End-Stage Liver Disease

There are very specific instances in which giving glutamine to a sick individual would not be indicated. Individuals who have severe cirrhosis of the liver, Reye's syndrome, or another metabolic disorder that can lead to an accumulation of ammonia in the blood are at an increased risk for encephalopathy or coma. The basic problem is an inability to clear the body of excess nitrogen, which is converted to ammonia and ultimately causes brain swelling and brain-cell death. When the liver is severely damaged or when hepatic coma is imminent, glutamine is not effective and would cause only further damage to the brain.

glutamine, while the remainder were given the standard glutamine-free solution. Most of the animals given the glutamine survived, whereas about 50 percent of the other group died. When the liver tissue was analyzed, it was found that the glutathione stores had been preserved or even increased in the animals that had been given glutamine but were markedly lower in the group that had been given the standard solution.

The drug 5-FU has many effects on the body. Because of the diffuse nature of the injury to tissues and cells caused by 5-FU, the investigators were unsure whether the glutamine contributed directly to the outcome by helping to make more glutathione and protecting the liver or whether the animals receiving the glutamine "just did better," which in itself preserves liver glutathione. The investigators therefore embarked on a second study, this one using acetaminophen, which in toxic doses *directly* depletes liver glutathione stores and causes severe liver failure. In this study, the glutamine-treated animals showed almost no mortality, had better hepatic glutathione stores, and displayed marked improvement in liver function.[11]

The investigators then wondered if this treatment could be applied to cases of infection or inflammation of the liver. They used an infection/sepsis model in experimental animals that activates the body's own cells to make large amounts of free radicals.[12,13] The injury usually causes severe inflammation in the liver, resulting in cell injury and death. Glutamine was given intravenously in total parenteral nutrition to some animals, and the results were compared with those of animals given TPN solutions that did not contain glutamine. Once again, the degree of liver injury was markedly reduced in the animals given glutamine. The intravenous diet that contained glutamine caused reduced inflammation, production of fewer inflammatory chemicals, less cell death and injury, and better liver function. Hepatic glutathione levels were also well preserved in the glutamine-treated group.

Thus, studies utilizing glutamine-containing diets have shown that glutamine protects the liver during toxic chemotherapy, during acetaminophen toxicity, and following a severe inflammatory injury to the liver. All of these effects may be due to glutamine's ability to support the antioxidant system of the liver. It is no wonder that many forward-thinking physicians are now using glutamine-containing solutions to treat patients following acetaminophen ingestion or with other forms of acute liver injury.

But there are other common ways in which free radicals are formed in the body. Think of people who "love the sun" or "love to smoke."[8] The free-radical damage they cause with these habits is visible on their thick, wrinkled skin. Other free-radical exposure occurs through air pollution, dry cleaning solutions, chemotherapy, or excessive alcohol consumption. Is glutamine the panacea for these damaging substances? The answer is still not clear. But if the acetaminophen overdose and recovery is a good indicator of how glutamine helps the body form glutathione, the answer may be the affirmative.

Glutamine and glutathione are currently being studied as important components of the treatment of people with AIDS. The initial work on glutathione and AIDS was done by Dr. Alton Meister's group at Cornell University.[14] Dr. Meister put the human immunodeficiency virus (HIV), the virus that causes AIDS, in a culture dish and found that the addition of glutathione prevented its growth. There are numerous studies underway in which researchers are giving forms of glutathione to people who are HIV-positive in an attempt to improve their conditions. Could glutamine help maintain and support glutathione levels in HIV-positive individuals?[15] Although the answers are not completely known, some compelling evidence links the provision of supplemental glutamine with glutathione levels. In addition to the studies already mentioned, another investigation was done in which researchers gave experimental animals a special enzyme that depletes glutathione in the body.[13] Then they gave the animals either a TPN solution that did not contain glutamine or a TPN solution that did contain glutamine. In the animals given a glutamine-containing solution, the level of glutathione rose, and the animals lived. Moreover, the animals' white blood cells were able to function better and could divide and kill bacteria. The other animals died from liver damage and immunosuppression because they had lost glutathione.

This may be the first step in the connection between glutamine and glutathione in AIDS patients. The liver is desperately trying to rid the body of the AIDS virus, but unless there is a steady supply of the right kind of precursor (preliminary substance), glutathione becomes depleted and the liver cannot do its job as effectively.[16,17] Glutamine is one of the major precursors of glutathione production.[11,18] If glutamine can help the body make more glutathione, perhaps glutathione can help the body fight HIV.

FATTY LIVERS AND GLUTAMINE

Glutamine has another important effect on the liver. The last decade has witnessed the increased use of very low calorie diets (VLCDs), in which individuals eat or drink only about 500 to 800 calories per day. We know that individuals on such diets develop fatty livers because when fat is being liberated from fat cells, the natural response of a liver near starvation is to accumulate the fat. As we mentioned previously, the liver is a great magician, converting sugar and fat into various other substances. However, when there is an overabundance of calories, or when there is partial fasting and rapid liberation of fat from the body's fat stores (as occurs with liquid weight-loss diets or severely restricted weight-loss diets), the liver becomes choked with fat. Such a liver is called a fatty liver; the liver has taken up too much fat and cannot get rid of it. The liver becomes enlarged—it gets fat—and shows signs of dysfunction when liver function tests are performed. If the liver is examined under a microscope, it is seen as clogged with fat.

Fat accumulation by the liver accompanies a variety of illnesses and diets. It is also related to ingestion of certain drugs and to excess alcohol use and is commonly associated with high-carbohydrate intravenous feedings. Scientists have discovered, however, that the formation of a fatty liver can be prevented with the addition of glutamine to the diet. In animals given intravenous feedings with and without glutamine, there was a marked difference in the livers. The animals given the standard glutamine-free diet had big livers, full of water and fat. The animals given glutamine had livers much the same size as the animals fed a normal diet of oral animal chow.[19] The fat and protein contents were also much more normal in their livers. It is difficult to know how a fatty liver contributes to overall disease. Clearly, under normal conditions, the liver should not be swollen nor contain a large quantity of fat. Therefore, this is another example of how the modification of diet with glutamine may aid in maintaining the health and function of the liver and thus contribute to the health and vitality of the individual.

Another negative side effect of high-carbohydrate intravenous feeding occurs in the liver's companion organ, the gallbladder. It is well known that people on total parenteral nutrition develop gallstones. Ultimately, some patients require gallbladder surgery. Scientists recently prevented the formation of gallstones in animals on TPN by supplementing the TPN feeding with glutamine.[20]

GLUTAMINE USE VERSUS GLUTATHIONE USE

Glutamine and the liver go hand in hand, in part because of glutamine's role in the formation of antioxidants. Newspapers and health magazines are filled with infor-

mation about antioxidants and their importance in protecting the body's tissues. Scientists feel that the positive
effects of glutamine shown in studies are due to the role
glutamine plays in the formation of the antioxidant glutathione. Glutathione's use as an antioxidant has been
described in numerous scientific articles, but rarely is
glutamine mentioned. Unfortunately, glutathione and
glutathione-like compounds are very expensive to take
orally. Not so with glutamine. Glutamine is relatively
inexpensive and readily available, and it greatly enhances the body's production of glutathione. Moreover,
people who take glutamine for its antioxidant properties
also receive the muscle-building, gastrointestinal, and
other immunological benefits of this important amino
acid.

5

Strengthening the Immune System

The study of bone-marrow transplant patients had gone on for two and a half years. None of the clinicians had known who had received the TPN solutions free of glutamine and who had received the solutions containing glutamine. As the young investigator tabulated the final data, he found that a large number of patients had suffered severe complications of infection, whereas others had seemed to be totally unaffected. Then it dawned on him—almost all of the patients who had been free of infection during their recovery from radiation and chemotherapy had received glutamine.

If you were in a stressful state of health, would you take an inexpensive, tasteless powder on a daily basis and reduce your susceptibility to infection? To put it another way, how would you like to be able to stay in as healthy a state as possible?

Some very sophisticated research has shown that taking the amino acid glutamine prevents major infections in patients who, because of treatment for illness, have almost completely destroyed immune systems. Before we look at this research, however, let us first consider what the immune system is and what effect it has in our bodies.

THE IMMUNE SYSTEM—A BRIEF REVIEW

Simply put, the immune system protects us from the external dangers of viruses, bacteria, toxic particles, and cancerous and other foreign cells. Our skin is the first protective barrier we have. Although it is not usually thought of as part of the immune system, it is the body's first line of defense in keeping out dangerous substances. It is a true physical defense against the outside world.

Inside the body, the intestines are another barrier. Imagine an infant picking up grass and soil, and eating them. The parents might be concerned, but this load of germs and indigestible grass would not be harmful to the child because most of the material would be neutralized in the bowel and expelled in the feces as part of the body's defense system.

The next barrier in the body is at the cellular level. Until the last thirty years, our knowledge of the molecular aspects of the immune system was fairly limited. The first bits of information came from observations of blood cells under the microscope. From the difference in the shape (morphology), color, and size of the various blood cells, scientists have been able to differentiate the defense cells (white blood cells) from the oxygen-carrying cells (red blood cells).

All white blood cells evolve from stem cells. The stem cells first divide into two major groups, myeloid and lymphoid precursors. The precursors in turn subdivide into specific white blood cell lines. Another system for classifying white cells is based on their physical characteristics and on the color they take up when they are stained with a certain kind of dye. Under this system, stained white cells are divided into neutrophils, monocytes, eosinophils, basophils, mast cells, lymphocytes, and plasma cells.

When you look through the microscope at stained white cells, you see an impressive array of colors. The eosinophils are easily distinguished from the basophils, which are easily distinguished from the neutrophils. The lymphocytes are far less impressive to look at. A lymphocyte is a small cell with a very large, dark nucleus, which is basically the brain of the cell. Given the large size of the nucleus in the lymphocyte, it is no wonder that the lymphocytes are the brains of the entire immune system. Once the lymphocytes detect a foreign substance, they initiate the immune response, which enables all the other immune cells to move into action.

As more has been learned about the white blood cells, their specific functions have been clarified. What has most amazed scientists is that all of these cells work in concert, each relaying messages to the others. In terms of complexity, the immune system is on a par with the nervous system, but it has an advantage over the nervous system in that its cells are not fixed to specific locations. The cells of the immune system are mobile and can migrate anywhere in the body, relaying messages among one another and receiving them from the outside world.

A critical ability of the immune system is recognizing "self" from "nonself." If the mission of an immune cell is to "seek and destroy," it must be able to detect when something is foreign to the body as opposed to a part of the body before it discharges its lethal chemicals. If an immune cell could not tell the difference between its own host (the body) and foreign organisms or particles, it would destroy the very tissues it was meant to protect.

The body is able to distinguish self from nonself by a special marker on the surface of all self cells. It is as if each self cell waves a flag announcing that it is self in order not to be destroyed.

The cells known as phagocytes (eating cells) eat up all foreign and worn-out matter in the body. As bacteria, foreign particles, worn-out cells, or cancer cells are detected in the body, the phagocytes wrap themselves around the culprits, engulf them, secrete powerful chemicals (free radicals and cytokines), and destroy the invaders. Some phagocytes are also able to recognize specific characteristics of foreign substances and to pass on the information to other defense cells, so that the other defense cells will be prepared to destroy the invaders if they slip by.

The main homes of the lymphocytes within the body are discrete areas known as the lymph nodes. In the neck, these are the areas that swell when you have a sore throat. These lymph nodes are attached to lymphatic vessels, which course throughout the body. There are also other centers of lymphoid tissue—the thymus gland in the chest, the spleen in the abdomen, the Kupffer's cells in the liver, the Peyer's patches around the intestines, the bone marrow, the tonsils and adenoids, and of course the cells circulating in the blood vessels.

All blood cells originate in the bone marrow. Since cells of the immune system are found in the blood, they too originate in the bone marrow. Some lymphocytes mature there and are referred to as B-cells. Others, called T-cells, migrate to the thymus and complete their maturation there. T-cells, also called T-lymphocytes or CD cells, can be divided into many subcategories according to their functions. T-helper cells help other T-lymphocytes and the B-lymphocytes do their job of destroying foreign substances. When the job is finished, T-suppressor cells begin to work to shut down the process. Cytotoxic T-cells are specialized killer cells that rid the body of cancer cells or cells infected with a virus.

B-lymphocytes are specialized to produce specific molecular compounds known as antibodies. Antibody molecules attach to foreign substances (antigens) and destroy them. Each B-lymphocyte recognizes only one antigen; it does so with the aid of T-helper cells. Once a B-cell is stimulated, it transforms itself into a plasma cell, which is like a factory for antibodies. Each plasma cell can discharge millions of antibodies into the bloodstream to find and destroy invading organisms. The antibodies produced by the plasma cells and B-lymphocytes are known as immunoglobulins (proteins composed of a characteristic pattern).

There are basically five main categories of immunoglobulins—IgA, IgD, IgE, IgG, and IgM. (*Ig* means "immunoglobulin.") Each plays a specific role in the mounting of the immune system's response. For example, IgA is found in body fluids such as the mucus in the respiratory tract or gastrointestinal tract. It attacks foreign substances that enter the body from the outside world as we breathe or eat various foods.

When an animal or human is unable to mount an immune response, he or she is said to be immunosuppressed. The most well-known state of immunosuppression is seen in people who have HIV, which leads to AIDS. People with HIV are extremely susceptible to bacterial and viral illnesses. Immunosuppression is also seen in other, less dramatic forms, such as in patients with chronic lung disease, in individuals on steroids such as prednisone, in people undergoing chemotherapy or radiation, and even in athletes who overtrain.

THE ROLE OF GLUTAMINE IN THE IMMUNE SYSTEM

What is glutamine's role in the immune system? In the 1950s, Dr. Harry Eagle at the National Institutes of Health realized that to grow immune cells in a glass dish (cell culture), glutamine was absolutely essential.[1,2] No glutamine, no growth. The implications of this work were lost to scientists for many years.

Dr. Eagle's cell culture work has been expanded at Oxford University in England by Dr. Eric Newsholme.[3] Dr. Newsholme found that if the amount of glutamine in a cell culture is reduced, the lymphocytes do not divide as they should. If the amount of glutamine is increased to normal, the lymphocytes again proliferate. Phagocytes, the eating cells, are best able to do their job of engulfing and destroying foreign material when glutamine is in the culture media. When the glutamine is reduced, the phagocytes become far less efficient.

Dr. John Alverdy, working at the University of Illinois, used laboratory animals to try to understand the relationship of glutamine to the immune system.[4] He found that if he gave a group of animals the same kind of intravenous nutrient solution that is standard in all hospitals in the world, their intestinal lymph nodes filled with bacteria. If he gave a similar group of animals the same solution but with glutamine added, there was minimal bacteria in the lymph nodes. He discovered that the antibody IgA, which is the one that keeps bacteria from entering the body, was at a high level in the animals that had received the added glutamine. It seems that glutamine helps the immune system do its job more effectively.

In another experiment, scientists gave laboratory animals massive doses of bacteria to make them septic.[5]

Glutamine consumption was measured in various tissues, and in the animals with sepsis, the lymphoid tissue was found to be using glutamine at an enhanced rate. This finding is hardly surprising, given that glutamine is required for lymphocytes and phagocytes to grow in cell culture and that lymphocytes multiply rapidly when called on to rid the body of foreign substances. In a different study, animals were again given enough bacteria to make them septic, but this time, some of the animals were given glutamine for one week before being given the massive doses of bacteria.[6] Only three of thirty-eight animals died in the glutamine group. Twenty-one of thirty-eight animals died in the non-glutamine group. Glutamine helped the animals fight infection.

A unique study was done by Dr. A. Baskerville in Britain that again reinforced the important role of glutamine in the body's defense system.[7] Dr. Baskerville gave several species of animals doses of an enzyme that has the capacity to use up completely all glutamine in the bloodstream. The animals rapidly became immunosuppressed and were unable to mount an immune response. They developed severe inflammation of the colon (necrotizing enterocolitis) and eventually died.

Most people are aware that chemotherapy has many inherent risks. One of the greatest risks is that this life-saving therapy can damage the immune cells and the bone marrow, where all the immune cells are produced, rendering the body susceptible to infection. Following chemotherapy, the intestinal barrier, which depends in part on the immune cells located there, becomes permeable, or leaky, thus allowing bacteria to get into the bloodstream and cause sepsis. A group of scientists replicated this situation in laboratory animals and then used glutamine to see if the effects could be reversed.[8] The animals were given chemotherapy. Those that were given the standard intravenous nutrient solution died soon after the chemotherapy began. Among the animals given the nutrient solution containing glutamine, however, there was a significantly reduced rate of death. The effect that the glutamine had on the intestine was thought to be twofold— it enhanced the ability of the immune cells to do their job of seeking and destroying the bacteria, and it prevented the intestine from becoming thin by enhancing the repair of damaged sites, thereby preventing the bacteria from passing through.

We have looked at numerous studies that indicate how important glutamine is as an immune enhancer in cell culture and in animals. Now let us address the glutamine information as it relates to humans.

IMMUNOSUPPRESSION AND GLUTAMINE

One study by Dr. Newsholme extrapolated cell culture data to humans. Dr. Newsholme found that when human lymphocytes are grown in a culture dish, if the amount of glutamine in the cell culture is decreased, the rate at which the lymphocytes replicate is also decreased, as is the ability of the phagocytes to phagocytose, that is, to destroy foreign matter. In patients with major body burns, the muscles are not able to manufacture enough glutamine for the blood levels to remain stable. Dr. Newsholme found that the concentration of glutamine in the blood of burn patients was 58-percent lower than was the concentration of glutamine in the blood of normal, non-burned individuals.[9] It remained low for twenty-one days. It is speculated that the decrease in blood glutamine may be a critical factor in the immunosuppression that occurs in patients who have suffered a major burn.

A study at the Shriners Burns Institute in Cincinnati confirmed that blood glutamine levels are low in patients who have been burned.[10] The patients in this study had burns covering from 10 to 81 percent of their body surfaces. Low glutamine levels were found in 73 percent of the patients, who for the study were divided into three groups. In addition to their regular tube feedings, the patients were given small amounts of glutamine added in incremental doses. The largest dose was only 6,000 milligrams (6 grams) per liter of fluid given. In all three groups, blood glutamine levels remained significantly below normal limits for the duration of the study. Additionally, it was found that the lower the glutamine level was, the more likely it was for the patient to have an infection of the burn wound. These data are consistent with Dr. Newsholme's theory that low blood glutamine may be a critical factor in the immunosuppression seen in burn patients.

AUTOIMMUNE DISEASES AND GLUTAMINE

Another fascinating role for glutamine is in an immune system that seems to be overactive—the body attacks itself and destroys its own tissue. These types of diseases are referred to collectively as autoimmune disease, for example, rheumatoid arthritis, systemic lupus erythematosis, polymyositis, and scleroderma. Frequently, people with these illnesses are described as having a "revved up" immune system and are often treated with a stress hormone, prednisone, which dampens the immune response; if the therapy is successful, the disease is held in check. The down side of this therapy, however, is that prednisone causes loss of muscle mass, loss of bone (osteoporosis), and increased vulnerability to infection.

Although autoimmune diseases can be controlled with immune-suppressive drugs, this control comes at a high price to the patient. Scientists have spent decades trying to understand why the immune system turns against itself. The answer to the problem may lie in a group of chemicals in the body known as cytokines. These chemicals are released from white blood cells such as macrophages. Tumor necrosis factor (TNF) and inter-leukin-6 (IL-6) are two examples of cytokines, but the specific names are not as important as the general reactions that they elicit in the body. In general, a small production of cytokines is a positive thing because cytokines attract immune cells to an injury or inflammation, enhance immune-cell proliferation and function, and orchestrate the healing and repair of the injured or inflamed area. But too great a volume of cytokines is extremely destructive. A cascade of reactions takes place, producing a greater and greater production of cytokines, all of which cause inflammation and pain, and destroy the body's tissue. For the last ten years, these compounds have excited medical researchers and deepened their understanding of what happens in the body during illness, injury, infection, stress, and autoimmune disease. This knowledge is also important for recognizing how to turn the cytokine system on and off.

What does this mean for individuals who have autoimmune diseases? People with an autoimmune disease like rheumatoid arthritis, even if it is under excellent control, have a 50-percent greater production of cytokines in the circulating cells of their blood than do people without an autoimmune disease. In the presence of these excess cytokines, their basic caloric needs are greater, but because of their high levels of cytokines, they have decreased appetites. They have chronic pain at the site of inflammation, and they have increased muscle wasting from cytokine production, from the steroids used to treat their conditions, from lack of exercise because of pain, and from lack of dietary intake because of decreased appetites.

It has now been discovered that glutamine, by contributing to the production of glutathione, is able to stop the production of cytokines when they are being synthesized in amounts harmful to the body.[11] Furthermore, glutamine enhances the growth of muscle and prevents further breakdown of muscle in animals and humans exposed to stress hormones like prednisone. Any patient on prednisone will readily testify to the muscle wasting and fatigue that accompany the therapy. Glutamine could be of real benefit to people with autoimmune diseases who are on steroids.

THE BONE-MARROW TRANSPLANT STUDY

Perhaps the most significant human study to date on the positive effects of glutamine on the immune system was done at a Harvard teaching hospital.[12] The patients studied were virtually without an immune system. They were bone-marrow transplant patients who had had cancer of the blood cells, or leukemia. If an individual who has leukemia is not treated, or if he or she does not respond to conventional cancer treatment, the leukemia cells proliferate to such an extent that the person dies. There was no effective treatment until not so many years ago.

However, a recent treatment, although fraught with its own risks, can be a lifesaving measure. The patient is matched up with a healthy person whose immune system is similar to the patient's and who is willing to donate some bone marrow. The patient is given massive doses of chemotherapy to kill the cancer. This is followed by total body irradiation, which completely destroys the patient's bone marrow, causes extensive injury to the intestines, makes all of the patient's hair fall out, and may render the patient sterile. Then the donor bone marrow is transplanted into the patient's body. The patient is kept in a germ-free room for as long as two months, until it becomes clear whether his or her body will accept the healthy bone marrow.

Up to the time the new immune system takes over, the patient is completely vulnerable to any infection that might pass through the environment. Even though the rooms are "germ-free," it is inevitable that some bacteria or viruses will get through. In fact, more than 80 percent of bone-marrow transplant patients are treated with antibiotics for some type of infection during the recovery period.

The glutamine study at Harvard was designed to answer many questions, one of which was: Can glutamine prevent or diminish infection in patients undergoing bone-marrow transplantation? The study involved forty-five patients, twenty-four of whom were given 40 grams (40,000 milligrams) of glutamine daily in small doses over twenty-four hours. Twenty-one patients were not given any glutamine, and all but one developed growth of bacteria in the throat, intestinal tract, or blood during the study period. Of the patients who were given glutamine, twelve remained completely free of bacteria during the same time period. This is statistically significant. The other surprising finding was that the patients who had been given glutamine were discharged from the hospital seven days earlier than were the patients who had not been given glutamine. The Harvard study has been replicated at the University of Kansas Medical Center under the direction of Dr. Paul Schloerb.[13] Dr. Schloerb's data confirm the early discharge from the hospital of the patients who had been given glutamine.

It is quite amazing to think that one amino acid, when included in the diet at a time of maximal immune vulnerability, can so greatly increase the ability of the body to fight infection. It is even more amazing to realize that this amino acid, glutamine, was previously thought to be nonessential and even today is not found in most nutrient solutions in hospitals.

A fact well known to doctors but not necessarily to the public is that a hospital is a terrible place in which to acquire an infection. Because so many antibiotics are used in hospitals, organisms like bacteria become increasingly resistant to standard, well-known antibiotics. Doctors must use increasingly more powerful antibiotics to win the battle against the organisms. It is now becoming obvious how important glutamine is in helping the immune system fight infections. The day is not too far off when all patients entering hospitals will be given glutamine-containing solutions or powdered glutamine to keep their immune systems strong and ready to fight hospital-acquired infections.

AIDS PATIENTS AND GLUTAMINE

What would be the effect of supplemental glutamine on AIDS patients, who also have limited immune function? Some work is currently in progress at Harvard University to examine that issue. As previously mentioned, when someone has suffered an extreme injury, such as a major burn, the muscle can no longer keep the blood levels of glutamine constant. In fact, levels in burn patients fall by 58 percent.[9] The people being studied at Harvard have the AIDS virus but no outward manifestations of AIDS. For the most part, they are normal, functioning individuals, except that they are HIV-positive. However, their glutamine levels are below normal, even though they are not ill with the AIDS disease.[14,15] Somehow, their bodies are rapidly using up glutamine beyond what their muscles can supply. Furthermore, a group of these individuals was given 20 grams (20,000 milligrams, or 5 teaspoons) of glutamine per day (in small doses over twenty-four hours) for one month and their blood levels of glutamine still did not rise.

These findings are surprising to the medical researchers. The individuals with HIV have no outward appearance of immune dysfunction, yet they are low in blood glutamine and 20 grams of glutamine per day did not raise their blood levels. This means that their muscle levels of glutamine are extremely low, which impairs the muscle's ability to synthesize more muscles, perhaps accounting for some of the muscle wasting of which HIV-positive patients complain. As the studies continue, the patients are now receiving 40 grams (40,000 milligrams, or 10 teaspoons) of glutamine per day, an amount equal to that which helped the bone-marrow transplant patients fight infection.

ATHLETES AND GLUTAMINE

Another area of research centers on potential uses of glutamine in healthy people. Dr. Newsholme at Oxford has proposed a theory in which the immune system, the brain, and the skeletal muscle are linked to the fatigue noted in athletes.[16] There are thought to be several main metabolic causes for fatigue in athletes that also may be why an athlete sometimes cannot give the final spurt needed to win an event. One of these metabolic causes is an increase in the ratio of the amino acid tryptophan to the branched chain amino acids in the human body during or immediately after exercise. This increased ratio allows tryptophan to enter the brain, where it is converted to the neurotransmitter 5-hydroxytryptamine (5-HT, also known as serotonin). This neurotransmitter appears to be responsible for fatigue in humans and animals.

Glutamine enters the picture because the immune system is highly dependent on it for the maintenance of its cells. When the immune cells need to replicate, they need even more glutamine. The main supplier of glutamine to the body is muscle. Unfortunately, when an individual "overtrains," muscles are unable to supply adequate glutamine, and blood glutamine levels fall. One study has shown that in overtrained athletes, glutamine levels were 9 percent below normal. This fall in glutamine may occur because the body pulls glutamine from the muscles to help the kidneys clear acid buildup (see page 19), so there is less glutamine available for the immune system.[17] This may be why highly trained athletes are known to develop more than the normal amount of infectious diseases like upper respiratory tract infections and to suffer delayed healing of wounds. A little exercise enhances the immune system, but too much decreases blood glutamine levels, promotes fatigue, and prevents the immune system from responding appropriately to an infectious insult.

Our knowledge of glutamine and the immune system in humans is expanding rapidly. The next few years will bring tremendous advances in our understanding of the role glutamine plays in maintaining health and preventing disease.

6

Helping Against Depression, Anger, and Fatigue

Sick people who added glutamine to their diets felt less depressed, less angry, and less fatigued. Alcoholics who added glutamine to their diets drank less alcohol. Glutamine increased the intelligence quotients (IQs) of both mentally impaired and normal children; in normal adults, glutamine was associated with improved problem-solving ability. Although information on glutamine's role in the brain and central nervous system is sparse, preliminary findings are certainly exciting.

A book on glutamine would not be complete without at least a brief review of glutamine's role in the brain and nervous system. Unfortunately, the information on this subject is sparse at best because it is difficult to develop sound studies on brain metabolism.

THE BRAIN AND GLUTAMINE

The same way that glutamine is the most abundant free amino acid in blood plasma, so it is the most abundant free amino acid in cerebrospinal fluid (the fluid surrounding the spine and brain).[1] Glutamine is the precursor of two important neurotransmitters—the excitatory neurotransmitter glutamic acid (glutamate) and the inhibitory neurotransmitter gamma amino butyric acid (GABA).[2] Neurotransmitters are substances that are deposited at the end of a nerve cell and picked up by the next nerve cell in order to send nerve impulses throughout the brain and the nervous tissue in the body. Because of the large amount of glutamine in the nervous system, glutamine is thought to be the storage element or precursor for glutamate and GABA.

One of the initial studies on glutamine's role in the brain was related to its potential for curbing addiction to alcohol.[3] Nineteen laboratory animals were given drinking fluid composed of alcohol and water. The animals that consumed the most alcohol were selected to receive glutamine. The selected animals were given 100 milligrams of glutamine per day orally, as well as unlimited access to alcohol. It was found that as long as the animals were receiving the glutamine, they drank an average of only 65 percent of their previous alcohol intake. Once glutamine was no longer given, the animals' intake returned to— and even exceeded—the previous high levels.

Similar research was then done with humans, but the dose of glutamine was only 1,000 milligrams.[4] The researchers noticed a positive response, but only a small one. This result, however, should probably be looked at with some skepticism because the dose of glutamine was so small. The average weight of the laboratory animals in the first study was 200 grams, whereas the weight of an average-sized man is 70 kilograms, or 70,000 grams. To be comparable to the dose given to the laboratory animals, the dose given to the humans should have been closer to 35,000 milligrams (35 grams, or approximately 9 teaspoons) per day.

In other studies, glutamine was given along with cytidine and uridine to individuals who had recovered from a stroke but were still exhibiting symptoms such as mild depression or compromise of attention, intelligence, or memory.[5] Electroencephalographs (EEGs) were used to measure improvement. The patterns on the EEGs showed general improvement in the electrical energy toward a more normal pattern when the patients were taking glutamine.

Glutamine was also tested on intellectually impaired children.[6] It was found to increase the IQs of the mentally impaired children, who were compared to a control group tested in a similar fashion.

Small quantities of glutamine (250 to 1,000 milligrams per day) were given to individuals with depression.[7] The results revealed clear antidepressive properties, and in the words of the researcher who conducted the study, a "vital" level of mental mood had been reached.

Having briefly discussed the results of studies in which very small doses of glutamine were given (usually in the range of hundreds of milligrams), we must now look at the results of more recent studies in which volunteers and patients were given not 250 milligrams daily nor 1,000 milligrams daily but 20,000 to 60,000 milligrams per day—quite a large difference. The purpose of one of the first studies was to determine the safety of such large quantities of glutamine.[8] As previously mentioned, glutamine is a precursor of two very significant neurotransmitters in the brain, glutamate and GABA. The researchers wanted to make sure that giving doses as high as 40,000 or 60,000 milligrams to normal volunteers would not cause coma, seizure, or any other adverse effects. The safety of large doses in normal individuals had to be established before any large dose of glutamine could be given to an individual with an illness. The question of whether glutamine is safe in large doses, at least in normal volunteers, was answered with an emphatic yes. In fact, testing on volunteers revealed that the ability to solve problems on the continuous performance test (CPT) improved while glutamine was taken, even though the study period was only five days. Perhaps a longer time on glutamine would have shown even greater improvement.

Another group of patients who underwent psychological testing while taking glutamine was the bone-marrow transplant patients (who were discussed in detail in Chapter 5).[9,10] Half of the patients were on glutamine at doses of up to 40,000 milligrams daily. The other half of the patients were given no glutamine. All of the patients took a simple, self-administered psychological test that asked about symptoms related to their moods since a person's sense of well-being and attitude toward illness can significantly affect the outcome of therapy.

The patients who received glutamine were statistically more "vigorous" than those who did not. The glutamine group showed improvement in other areas, too. They felt less angry and less fatigued than the patients who were not given glutamine. If you consider the sense of depression that accompanies most illnesses, you might ask whether such depression could be diminished through the use of glutamine. The answer to this very important question could possibly affect all of us.

PAIN AND GLUTAMINE

The perception of pain is a significant function of the nervous system and is one of the body's most important defense mechanisms. Pain lets us know that something is wrong. There are times, however, when we would like to diminish our perception of pain—for example, after an operation. A series of studies revealed that glutamine provided analgesia (pain relief) to laboratory animals.[11] A standard technique was used to assess pain, and glutamine was found to be almost as effective as phenylbutazone (a compound similar to aspirin) in reducing reaction time to the painful stimulus. In this same series of studies, the amount of inflammation and edema (tissue swelling) was measured after the animals were injured, and the animals receiving glutamine had significantly lower levels.

Currently, glutamine is being used at the Brigham and Women's Hospital in Boston to heal mouth sores and ease the pain of cancer patients who have oral mucositis (inflammation of the skin of the mouth) from chemotherapy.[12] The patients are given 5 grams (5,000 milligrams, or 1¼ teaspoons) of glutamine six times a day. The powdered glutamine is dispensed in a small amount of water, which the patients swish around in their mouths for a few seconds and then swallow. The patients report that it has a soothing effect and relieves the pain of open sores. Of course, the greatest benefit is that the mouth sores associated with the chemotherapy soon disappear entirely.

Are there other ways in which glutamine can help ease the pain of cancer and the side effects of chemotherapy and radiation? The results of some exciting research in this area are presented in Chapter 7.

7

Fighting Cancer

The woman who came to see the surgeon had inflammatory breast cancer, one of the most difficult types of breast cancer to treat. Usually, patients must take many different kinds of chemotherapeutic drugs for this cancer. The surgeon had been studying glutamine's role in improving the results of chemotherapy, and she decided to include glutamine in this patient's chemotherapy regimen. The patient was given glutamine orally every day, and after she had been taking it for four days, she was also given high doses of one chemotherapeutic drug, methotrexate, once a week. She continued to take the glutamine daily. After three weeks, the surgeon operated on the woman to remove as much of the tumor as possible. To everyone's amazement, the woman had no trace of cancer remaining in her breast.

In Chapter 1, we learned that glutamine is absolutely essential for the growth of cells in laboratory cell culture dishes. This is true for all cell types, including cancer cells. Because cancer cells are also dependent on glutamine, doctors have been reluctant to give glutamine to patients who have cancer. An early exception was the important Harvard study of patients who required bone-marrow transplantation. New studies with humans are underway, although the results have not yet been published, and more human studies will no doubt be forthcoming because of the exciting data obtained from the following investigations using animals.[1,2]

TUMORS AND GLUTAMINE

In one of the first studies on cancer and glutamine, Dr. Wiley Souba and associates at the University of Florida at Gainesville implanted tumors in animals; these tumors were allowed to grow until they were 5 percent (small tumor) or 10 percent (large tumor) of the animals' weights.[1] The animals that were given glutamine supplementation did not have larger tumors or more actively growing tumors than did the nonsupplemented animals, although the proportion of tumor cells to host cells within the tumors was increased. Furthermore, the glutathione levels in the intestines were increased, which would naturally be a form of protection for animals.

A study by Dr. Suzanne Klimberg at the University of Arkansas took this idea even further.[3] Again, animals were implanted with tumors, and the tumors were allowed to grow for twenty-five days. The animals were divided into groups and either supplemented with glutamine or not. According to the results, tumor growth was similar in the animals given and not given glutamine. But in the animals that were subsequently given chemotherapy, there was a 45-percent reduction in tumor size in the animals that received glutamine. There was only a 25-percent reduction in tumor size in the animals that did not receive glutamine. Therefore, in some way, glutamine was able to enhance the ability of the chemotherapy to kill the tumors. Furthermore, the animals given chemotherapy but not glutamine had a 100-percent infection rate, whereas the glutamine-supplemented animals had only a 3-percent infection rate. There was also a statistically significant number of survivors among the glutamine-enriched animals as compared to the animals not given glutamine. All this information points to the important role glutamine may play for the millions of people who undergo chemotherapy every year.

Even more remarkable findings came out of another study in which cancerous tumors were implanted in animals.[4,5] The cancers were allowed to grow for twenty-three days, then the animals were divided into two groups—those who were given glutamine and those who were not. After two days of this special diet, all of the animals were given methotrexate, a chemotherapeutic drug. At twenty-four and forty-eight hours after the methotrexate was injected, the animals were tested, and those that were being given glutamine were found to have a significantly greater concentration of methotrexate in their tumors than did the animals on a glutamine-free diet. Moreover, the tumors were significantly smaller in the glutamine-treated animals. Another important finding was that the glutaminase levels in the tumors of the animals given glutamine were lower, indicating that because there was less tumor, there was not so great a need to utilize glutamine for its growth. This indicates that glutamine is important not only for its nutritional value but also because it has specific therapeutic value as a supplement to chemotherapy. We know that patients with cancer who can tolerate higher doses of chemother-

The patient had a tumor in his mouth, and he was being treated with both radiation and two kinds of chemotherapeutic drugs. This treatment usually makes patients very sick, but this patient was being given oral glutamine as well. At the start of the cancer treatment, while on glutamine, the patient actually gained weight. He never developed the mouth ulcers that this treatment usually causes, he continued to have salivary output (this treatment usually causes the salivary glands to shut down), and his cancer responded well to the therapy.

apy survive longer. The use of glutamine may allow for higher doses of chemotherapy because there are fewer toxic side effects, such as mucositis and diarrhea. Moreover, glutamine may have a direct role in preventing the growth of the cancer when it is used with other chemotherapeutic agents.

RADIATION INJURY AND GLUTAMINE

Dr. Klimberg then turned to the issue of radiation injury.[6] As with chemotherapy, radiation therapy kills cancerous tumor cells. Likewise, it destroys normal tissue, especially tissue that has a rapid turnover, such as the intestines and the mucosa of the mouth and esophagus. One of the worst complications following radiation therapy for cancer of the female organs is chronic radiation injury to the intestines with possible bowel obstruction. Dr. Klimberg and her associates wanted to see if they could prevent chronic radiation injury by using glutamine, and they developed a novel experiment to test their theory. In laboratory animals, they surgically brought down a loop of intestine and isolated it from the rest of the intestines.

All of the animals were treated with radiation to the isolated, normal loop of bowel. One group of animals was given glutamine prior to and during radiation; the other group received standard treatment. The glutamine-treated group had no evidence of radiation injury to the bowel; the other group revealed the usual radiation damage. This experiment proved so exciting that studies are now underway using glutamine supplementation in female patients who have gynecological cancers requiring radiation to the abdomen.

The ways in which glutamine enhances the tumor-killing properties of chemotherapy and protects against radiation-induced injury are felt to be due to the role it plays in the formation of the antioxidant glutathione. This role in glutathione production is also believed to be the reason for the positive effects that glutamine has on the liver and why, as we will see in Chapter 8, it may help older people to remain healthy and vigorous.

8

Making Aging Easier

It is never too late to change your lifestyle or general habits. Granted, people who have healthy habits early in life are more likely to continue to be healthy, but at no point in life are people unable to improve their bodies or their physical or mental health. A perfect example of what the body is capable of if it is well tuned is demonstrated by Norman Vaughan, who had a glacier named after him in the Antarctic. He and a team of explorers plan to climb to the summit of the 10,000-foot peak. This adventure is even more noteworthy because Mr. Vaughan is in his eighties. Each one of us has tremendous untapped potential.

Of all the age groups between birth and old age, those with the most diverse ranges of skills and levels of functioning are the older age groups. The process of aging is certainly not equal among people. We can all think of vigorous, robust eighty-year-olds and feeble, demented sixty-year-olds. Much of the aging process, or the degree to which we "show our age," is determined by nutrition, environment, and lifestyle. Although the nutritional needs of individuals change throughout life, many of the detrimental changes we see as we age are effects of lifestyle and life choices. This chapter will focus primarily on the aging process, with emphasis on nutrition, vitamins, and glutamine.

THE AGING POPULATION OF THE UNITED STATES

At the turn of the twentieth century, one out of every nine people in the United States was over age sixty-five. By the year 2000, one out of every four people—25 percent of the population—will be over that age. Life expectancy has also been increasing. In 1900, the average life expectancy was forty-seven years. For those born in the 1990s, the average life expectancy is seventy-four years. A part of this increased life expectancy can be attributed to improved nutrition, as well as to the development of antibiotics and other public health measures such as vaccinations and sanitation.

In general, population scientists assume that an increased life expectancy reflects the improved health status of a nation. This may not be so in the United States, however. It now appears that Americans are extending their lives but are living with more chronic and debilitating diseases than was done by recent past generations. That is, we are extending life quantity but not life quality. Many of the chronic diseases that people are developing are related to either inappropriate lifestyle or inadequate nutrition. By lifestyle, we mean the choices we make about how to live our lives. For example, a lifestyle choice is whether or not to smoke cigarettes, or whether or not to engage in a program of regular exercise.

For tens of thousands of years, the main function of humans was to forage and hunt for food. Suddenly, within three generations, we have gone from being muscular and trim from working in agriculture and at other manual labor to being sedentary and out of shape. As a result, approximately 40 percent of adult Americans are overweight or markedly obese. It appears that we no longer consider food to be a source of nutrients and a pathway to health. We seem to eat impulsively whatever we please. Unfortunately, this attitude is contributing to the rampant malnutrition (in the form of obesity) and nutritional disorders in the American population. (*Malnutrition* is defined as abnormal nutrition, and in the context of developed nations, it often refers to overnutrition, as in people who are overweight rather than undernourished.)

GOOD NUTRITION IN OLDER PEOPLE

Unfortunately, as we age and reduce our activity levels, our need for calories lessens. Because we have become accustomed to eating calorie-rich foods or consuming alcohol, however, we tend to continue such consumption and to eliminate from our diets the foods that provide the greatest amount of nutrients, such as fruits and vegetables. If we are not mindful of what we eat, the likelihood of our not getting appropriate nutrients also increases. This has a significant impact on how we age.

There are many causes of poor nutrition in individuals as they age. The primary causes include inadequate knowledge of nutrition, poverty, social isolation, depression, physical disability, and poor dentition. Secondary causes include gastrointestinal disorders and malabsorption, use of alcohol, chronic drug treatment, and illness.

We can look at the primary organ systems as they

specifically relate to nutritional problems in the elderly. The entire gastrointestinal tract, from the mouth to the colon, plays an important role. A whopping 50 percent of people over the age of sixty-five in the United States have lost all of their teeth.[1] Many of the remaining 50 percent have partial loss of teeth or periodontal disease (gum disease). Tooth loss or periodontal disease cause the act of chewing food to become difficult. Chewing is made even more difficult by poor dentition or dentures. If chewing is painful because of problems with the mouth or teeth, a person is less likely to eat raw fruits and vegetables, which contain far more nutrients than do mushy canned or cooked items. If the facial muscles are not used, they begin to atrophy, and eating becomes further impaired.

The tissue lining the mouth and tongue becomes thinner as people age.[2] This leads to increased susceptibility to the formation of mouth sores and ulcers, which can become infected. Older individuals often complain of dry mouth, which may be caused by the numerous medications they take or may come from mouth breathing, which is often seen in people with respiratory problems. Gastric secretions diminish with age, and physicians often detect gastric atrophy or gastritis in the elderly. Surprisingly (since ulcers are usually thought of as occurring in younger individuals), gastric ulcers actually increase with age, and complications such as bleeding frequently lead to death.[3,4] It is known that the elderly take more medications than do younger individuals. One group of medications frequently given to the elderly for the pain and swelling associated with arthritis is the NSAIDs. Unfortunately, in many people, these medications cause stomach ulceration, gastritis, or bleeding.

Aging is associated with atrophy of both the muscular

layer and mucosal lining of the small intestine. The large intestine frequently develops diverticular disease (inflammation of small outpouchings of the colon) and inflammatory bowel disease.[5]

The musculoskeletal system is also significantly affected by aging. It is well known that as individuals age, they lose muscle mass, known as lean body mass.[6,7] Even with continued active exercise, there appear to be changes in muscle mass with age. Muscle strength appears to be greatest at around age thirty and remains relatively constant until age fifty or sixty. After age sixty, the amount of strength a person has depends on many factors, including how active the person is and his or her state of health and nutrition.

In the United States, the fourth leading cause of death in the elderly is infection. Scientists have found that the immune system does not react as vigorously in older people as it does in the young. In other words, we are far more susceptible to pneumonias and other infections when we are in our sixties than when we are in our twenties.

LACK OF CORRECT NUTRIENTS IN OLDER PEOPLE

How does nutrition affect the aging process? Since the science of nutrition became a recognized field during the twentieth century, its primary emphasis has been to prevent conditions related to specific nutrient *deficiencies*. For example, the use of vitamin C has been encouraged to help maintain and promote strong, healthy tissue and prevent the development of scurvy. The manifestations of scurvy were first noted two centuries ago in sailors who developed bleeding gums and subsequently lost their teeth on long voyages. Eating fresh oranges and other citrus fruits was found to prevent these complications because the fruits contain vitamin C.

Deficiencies of the other vitamins lead to their own specific conditions or diseases. They include:

- *Vitamin A*. Night blindness, dryness of the conjunctiva and cornea, poor growth, and dry skin.
- *Vitamin B₁ (thiamine)*. Beriberi (nervous tingling, poor coordination, edema, heart changes, and weakness).
- *Vitamin B₂ (riboflavin)*. Ariboflavinosis (inflammation of the mouth and tongue, cracks at the corners of the mouth, and eye disorders).
- *Vitamin B₃ (niacin)*. Pellagra (diarrhea, inflammation of the skin, and dementia).
- *Vitamin B₆ (pyridoxine)*. Headache, anemia, convulsions, nausea, vomiting, skin flakiness, and tongue soreness.
- *Vitamin B₁₂ (cobalamin)*. Anemia and poor nerve function.
- *Vitamin C*. Scurvy (poor wound healing, pinpoint hemorrhages, bleeding of the gums, and edema).
- *Vitamin D*. Bone disorders.
- *Vitamin E*. Damage to red blood cells and destruction of nerves.
- *Vitamin K*. Hemorrhage.
- *Pantothenic acid*. Tingling in the hands, fatigue, headache, and nausea.
- *Biotin*. Inflammation of the skin, tongue soreness, anemia, and depression.
- *Folic acid*. Anemia, inflammation of the tongue, diarrhea, poor growth, and mental disorders.

The recommended dietary allowances (RDAs), which are frequently mentioned in articles on nutrition in newspapers and magazines, give the recommended amounts of the nutrients that should be ingested to prevent deficiencies in a normal person.[8] The amounts vary according to age and gender as well as pregnancy, although the labels on most food packages do not take this into account. Many leading nutritionists now consider the RDAs to be inadequate because evidence is accumulating that shows that certain nutrients, if taken in doses much higher than recommended, may actually *prevent* disease—which is much different than preventing a deficiency. This is causing tremendous excitement and controversy in the nutritional sciences. (For additional discussion of this subject, see page 90.)

During the 1970s and 1980s, researchers discovered many vitamin deficiencies in the United States among young and old alike, but they found that deficiencies were most prominent in older people. A survey in the Boston area revealed that approximately 65 percent of older people living on their own took in less than two-thirds of the RDAs for vitamins B_6, B_{12}, and D; folic acid; zinc; calcium; and chromium.[9] This is disgraceful when you consider that the United States is called the land of nutritional plenty. In fact, if the survey had revealed just how many older people took in 100 percent of the RDAs for those nutrients, the results would have looked even bleaker. One explanation is that as people age, they require fewer calories, and they often sacrifice nutritious foods in favor of more high-calorie, fatty foods and/or alcohol. Another part of the problem is that many elderly people do not have the resources or the energy to shop or cook for themselves.

Deficiencies can lead to numerous problems that complicate aging and in some cases accelerate the aging process.

An example is vitamin D deficiency leading to osteo-porosis (thinning of the bone structure).[10,11] Because vita-min D helps in the process of bone remodeling, a deficiency in vitamin D contributes to osteoporosis. In the northern latitudes of the United States and in Canada, people cannot synthesize adequate amounts of vitamin D in the skin be-cause the sun is not bright enough from October through April. Those who are intolerant of milk or who obtain their dairy products only in the form of yogurt or cheese are at risk for developing osteoporosis unless they take 400 inter-national units (IU) of supplemental vitamin D daily during the winter. For individuals who live closer to the equator, the problem is not as great, although now that people are encouraged to use sunscreens, the risk of vitamin D defi-ciency during the remaining months has increased.

When inadequate intake of nutrients is combined with poor absorption of nutrients from the gastrointestinal tract, significant problems with accelerated aging may occur. In a Boston survey, it was discovered that there was a 24- to 37-percent incidence of atrophic gastritis in individuals over sixty years of age.[12] This condition, which literally means "wasting of the stomach," is associated with decreased pro-duction of hydrochloric acid, which is a crucial ingredient for the digestion and absorption of foods and also for the absorp-tion of vitamin B_{12} and folic acid. In fact, absorption of these vitamins decreases by 50 percent with atrophic gastritis. If you will recall, the intake of B_{12} and folic acid in a similar population in the Boston area was also significantly reduced. What does this mean for the individual?

When B_{12} and folic acid levels are low, homocysteine levels are high.[13] Homocysteine is toxic to blood vessels and nerves. High levels are felt to be a significant risk factor for coronary heart disease, which is the number-one cause of death in the United States and accounts for an enormous amount of pain and suffering, to say nothing of its contribution to the aging process.

The current RDA for vitamin B_6 for the elderly is 2 milligrams per day. However, research at the U.S. Department of Agriculture (USDA) Human Nutrition Research Center on Aging at Tufts University in Boston revealed that the need of elderly individuals is for 3 milligrams per day.[14] We know that many people do not get even the recommended 2 milligrams per day, much less 3 milligrams. Without adequate B_6 stores, people are at increased risk for neuropsychiatric disorders and depression.[15,16] All of these conditions accelerate aging.

STRESS, OLDER PEOPLE, AND GLUTAMINE

In addition to the long-term effects that lifestyle and nutritional choices have on the body, the effects of stress also increase as a person ages. After a stress, the elderly body never fully recovers. A person loses weight in increments, dropping a few pounds with each major stressful encounter or complication associated with a chronic illness. Along with weight, nutrients necessary for health maintenance are lost. The person becomes underweight, malnourished, and susceptible to death from stresses as small as a bout of flu.

It is very sad indeed that in this country of abundance, some people are aging very rapidly because of inadequate intake of nutrients. Osteoporosis, heart disease, depression, diabetes, and blood disorders in some instances could be improved with adequate diet and, when indicated, with nutritional supplements.

ILLNESS, OLDER PEOPLE, AND GLUTAMINE

It is not prudent to eat an unbalanced diet and to then try

to compensate by taking vitamin pills and other nutritional supplements. However, in some situations, supplements are necessary. For example, it is not possible to eat food in large enough quantities to obtain the amounts of vitamins necessary to prevent certain diseases and ailments.

As discussed on page 87, scientists for several decades have studied ways to prevent deficiencies of nutrients, but until quite recently, they did not consider whether taking extra nutrients might prevent *illness.* A study by Dr. R.K. Chandra using a group of middle-class, apparently well-nourished, elderly Canadians revealed some surprising results.[17] Half of the group was given a multivitamin supplement with minerals, extra vitamin E, and beta-carotene. The other half was given a placebo. Over a period of one year, the participants reported when they had colds or other infection-related symptoms. Those taking the vitamin tablet reported less than half as many sick days during the year than did those taking the placebo. None of the participants had overt signs of nutritional deficiency, yet the people taking the vitamin supplement benefited greatly. Clinical investigators have now begun to appreciate more fully that the nutritional needs of different segments of the population vary significantly and that certain vitamins, particularly antioxidants, may be needed by some in greater-than-normal doses. For example, researchers at the USDA Human Nutrition Research Center on Aging gave some elderly people a vitamin E supplement and others a placebo.[18] Within one month, those taking the vitamin E supplement showed improved immunity, whereas those taking the placebo showed no change in immunity. This study was important, for none of the subjects was deficient in vitamin E; the study showed that *supplementation* of vitamin E to elderly individuals over and above the normal requirements enhanced immune function.

Since 1972, the Harvard School of Public Health has

been studying 121,700 female nurses in a project known as the Nurses' Health Study.[19] Every two years, the nurses answer questions about their lifestyles and medical histories. *The New England Journal of Medicine* (NEJM), perhaps the most prestigious medical journal in the world, reported that the study has so far revealed a 40-percent decrease in coronary heart disease and mortality in those women taking a vitamin E supplement for two years. In the same issue, it was reported that similar results were found by the Harvard researchers in a group of male health professionals. The study also noted a statistically significant protective effect from high carotene intake in men who smoked cigarettes.[20]

Another dramatic example is seen in smokers. An individual who smokes cigarettes has an increased need for vitamin C.[21,22] Do you know smokers whose doctors recommend that they eat foods high in vitamin C or take a vitamin C supplement? In most instances, the doctor encourages the patient to quit smoking but does not recommend a change in diet.

As many people age, they develop cataracts, which are caused by ultraviolet-induced damage to the eye's lens crystalins. People who have low vitamin C levels are at three to seven times higher risk for cataracts. In one study, people who took 300 milligrams of supplemental vitamin C per day increased their *protection* from cataracts four-fold.[23] Epidemiological research has recently shown that people who get at least 50 milligrams of vitamin C from food plus take a vitamin C supplement of at least 500 milligrams extend their life expectancy by an average of five years (seven years for men and three years for women).[24] Another study revealed that a 2-gram dose of vitamin C helped participants handle alcohol much better.[25]

Sunlight, smoking, and alcohol all produce oxygen free radical damage in the body. Vitamin C, along with vitamin E and beta-carotene, belongs to the class of vitamins known as antioxidants. They protect the body from free-radical damage.[26-31] Glutamine indirectly belongs to the class of compounds called antioxidants because of its role in producing glutathione, the most important free-radical scavenger in the body.

What are some of the usual age-related changes that may be improved through the use of glutamine? One that comes prominently to mind, although most people do not think of it as an ailment, is the loss of strength and slow recovery from illness. Other than lack of use, the most obvious reason muscle mass is lost is that the body needs to break down muscle to form glutamine for the immune and gastrointestinal systems. Unless a person actively works to rebuild muscle after an injury or illness, the muscle is lost forever—only to be replaced by wasted tissue or fat. It seems that the full muscles you had in youth are irretrievable, although this situation need not develop. If people would supply their bodies with glutamine in times of stress, they could prevent muscle from wasting away. Remember, blood glutamine levels must be normal before muscle can be regenerated.

Ironically, when a person exercises strenuously, as in vigorous weight training, oxygen free radicals are produced; these are cleared in part by glutathione. In addition, the body produces lactic acid and other harmful byproducts that need to be cleared through the kidney; this process also utilizes glutamine. Therefore, it seems that the positive process of restoring the body's strength and muscle has a negative side effect of producing harmful compounds. The way to minimize the effects of these negative compounds is to have enough antioxidants such as vitamins C and E, beta-carotene, and of course glutamine.

STRENGTH, OLDER PEOPLE, AND GLUTAMINE

Another age-related change is the deposition of fat at the expense of muscle and the unfortunate lessening (at least for most of us) of the number of calories required to sustain ourselves. As we age, our basal metabolic rate lowers, that is, the body needs fewer calories while lying in a restful (basal) state. Research is revealing that some of this decrease is not inevitable. It is secondary to loss of muscle mass, which occurs from lack of use as we age. Therefore, if people worked to maintain their strength or redevelop lost muscle through exercise, their caloric requirements would increase. Simply put, muscle burns calories; fat does not. What better reason to work at maintaining strength than to be able to enjoy your favorite treat without paying for it on the scale? More important, of course, is that a strong person is able to maintain independent living for far longer than a weak-bodied person can. Normal glutamine levels facilitate the deposition of muscle.

Numerous ailments that affect us as we age require medication. Often, people who never took medication find themselves consuming a handful of pills every day. Prominent among those medications are aspirin and the general class of NSAIDs, which include ibuprofen. As we age, we seem to have a lot more minor aches and pains, and our use of such medications increases while our stomachs seem to get upset more easily (perhaps from atrophic gastritis). Glutamine can protect the stomach somewhat from the gastritis caused by NSAIDs.

If we focus solely on the gastrointestinal system, from the mouth to the colon, it appears that glutamine could be of real benefit to the elderly. Two of the prime sites of glutamine utilization are the stomach and intestine, and glutamine is known to specifically prevent the atrophy associated with decreased food intake. Glutamine is also beneficial as an antiulcer remedy and is known to reduce pain and swelling after injury in an animal model. The muscular layer of the intestine thins as we age, and glutamine acts like a specific growth factor for muscle. All of these effects are enhanced by eating an adequate, well-balanced diet.

Another very important function of glutamine for people who must take a lot of medication is protection of the liver. As discussed in Chapter 4, the liver is the primary site of detoxification of all substances taken into the body. As people age, their capacity to detoxify many medications begins to wane. The reason for this is not completely understood. However, it is known that glutamine is the primary protector of the liver, as evidenced by its role in preventing damage from chemotherapeutic drugs and acetaminophen overdose (see page 45). Perhaps older people could handle medications better if their livers were able to function at the maximum, with the help of glutamine. Certainly, if someone is going to drink alcohol at the expense of more nutritious foods, glutamine can help to protect the liver from the resulting oxidative stress, that is, the free-radical production. In fact, in Japan, most people take a tablet containing glutamine when they know they will be drinking alcohol. This tablet is considered a hangover preventative.

Many people wish to take advantage of the latest research on antioxidants but are at a loss concerning what to take and how much. There are no recommended dosages of the vitamins to prevent deficiency states other than the RDAs. However, Dr. Chandra gave a multivitamin tablet with minerals daily to the people in his study with no untoward effects. In fact, the subjects taking the daily multivitamin-and-mineral tablet did much better than did the subjects taking the placebo. You might consider taking a high-stress multiple vitamin. In addition, you might supplement the multivitamin with the antioxidant vitamins, taking 500 to 1,000 milligrams of vitamin C, 100 to 400 IUs of vitamin E, and 15 milligrams of beta-carotene. However, you should do this only with the approval of your health-care provider. There is little risk to taking antioxidant vitamins, but there is considerable benefit in protecting the body from free-radical injury from air pollution, smoking, alcohol, ultraviolet light, and so on.

Unfortunately, when it comes to vitamins, more is not always better, particularly with vitamin A. It appears that as a person ages, his or her ability to handle this fat-soluble vitamin diminishes, and it is possible to get an overdose of vitamin A with doses very close to the RDA. Some people may be aware that beta-carotene is a precursor of vitamin A. However, the wonderful thing about beta-carotene is that it will not convert to vitamin A unless the body needs vitamin A.

9

Adding Glutamine to Your Life

Now that you have read all about glutamine, you may have some questions about how to use it. It must be emphasized again that glutamine should be used only under the supervision of a doctor. If your physician does not know about glutamine, we suggest that you give him or her this book to read. Glutamine is one of the truly remarkable discoveries in nutrition in the last fifty years, and your doctor needs to know about it.

So, before we sign off, let us discuss the details of how to add glutamine to your life. In this chapter, we will cover where to find glutamine, how to take it, how much to take, who should not take it, and its possible side effects.

SOURCES OF GLUTAMINE

Unfortunately, foods are not a good source of glutamine. As you might imagine, the foods highest in glutamine are muscle (meat, chicken, and fish) and eggs. In living animals, muscle has the capacity to make large quantities of glutamine in times of stress, but as a component of the protein in meat, glutamine makes up only 3 to 4 percent of the total amount of amino acids in the protein. Furthermore, heat rapidly denatures (inactivates) glutamine, so cooked meat, fish, chicken, and eggs contain even lower amounts of the amino acid than do the raw foods. The only practical way to add glutamine to the diet is to take supplements or to eat raw meat, fish, or eggs, the latter of which is of course *not* recommended—although the Japanese may be on to something in their fondness for *sashimi* (raw fish).

If it seems confusing that meat contains little glutamine, look at it as follows. Animals, including humans, have the capacity to respond dramatically to a stressful situation. The entire body reacts with something called the "fight or flight" response, during which all the free glutamine in the muscle is discharged into the bloodstream. The bloodstream carries the glutamine to the tissues that need it the most, such as the intestines, the immune system, and the liver. Muscle has the capacity to then produce additional glutamine by converting more muscle protein. After a relatively short time (several days), however, the muscle becomes quite wasted as the supply of glutamine becomes depleted. From a teleological point of view, animals were meant either to survive this major insult or to die. It is only because of modern medicine that individuals with major injuries or illnesses continue to live and to recover from their medical problems. We are discovering that in this setting of increased body stress, glutamine must be given to supplement the glutamine produced by the muscle and to prevent the muscle from wasting to the point where the individual is incapacitated and slow to recover.

Given that foods contain little glutamine, we must look elsewhere for a source. Glutamine is available as a nutritional supplement. Most health food stores carry glutamine tablets; however, the tablets for general consumers contain just 500 milligrams (0.5 gram) of glutamine. The amounts given to patients following surgery or bone-marrow transplantation range from 20,000 milligrams (20 grams) to 40,000 milligrams (40 grams) daily, given in small doses over twenty-four hours. One teaspoon of glutamine powder equals just 4,000 milligrams (4 grams).

Two companies sell nutritional products containing glutamine for use in specialized diets. Ross Pharmaceuticals distributes Alitraq, which contains 15.3 grams of glutamine in every 1,000 calories worth of product. Sandoz sells Vivonex, which contains 4.9 grams of glutamine in every 1,000 calories worth of product. The main manufacturer in the world of glutamine is the Ajinomoto Company, based in Japan. It sells glutamine to distributors, which then repackage it for sale to pharmaceutical and health food companies.

It is somewhat amazing that glutamine is not added to hospital food. Given the fact that most people are under physiological stress when they are in the hospital and that many are in a semistarvation state due to meals being withheld because of testing, it would be logical for glutamine to be added to all hospital food. This is not the case, however, although many hospitals are working to change this unfortunate practice by incorporating more glutamine into the intravenous fluids and diets of their patients. Within five to ten years, all hospitals will probably be utilizing this special nutrient to help prevent the infections to which so many patients are susceptible.

HOW TO TAKE GLUTAMINE

Although we know a fair amount about the benefits of glutamine, no one really knows for sure exactly how much glutamine should be taken. There is no single recommended dose of glutamine. Completely healthy individuals do not need to take any glutamine at all, and the amount needed by an individual who is under stress is still unknown. Some people take 1 to 2 teaspoons (4 to 8 grams) of glutamine per day if they feel they are at increased risk for stress because of dieting, heavy exercise, flu, diarrhea, an upper gastrointestinal problem such as a stomach ulcer, or intestinal bowel disease. In Japan, people who have ulcers take 1 gram of glutamine three times a day.

At a major Harvard teaching hospital in Boston, the dietitians act very quickly to provide glutamine to patients who need it. A patient with diarrhea is usually given a nutritional product containing glutamine. While the nutritional product is being brought up to full strength, the patient is also given supplemental glutamine powder mixed with water at a dose of 0.57 gram of glutamine for every 1 kilogram (2.2 pounds) of body weight. The results have been excellent. Adults with diarrhea who do not need a special nutritional drink are usually given 1 teaspoon of glutamine powder in water six times per day.

We do not yet know how much glutamine is too much. Glutamine safety studies have been done on healthy volunteers using doses of up to 0.75 gram of glutamine for each 1 kilogram of body weight.[1,2] No adverse effects have been noted at these levels. However, no one knows what would happen with higher doses, since such studies have not yet been done.

Always take glutamine at the doses recommended by your personal physician. Medical supervision is particularly important for individuals who are ill, particularly people with kidney disease or severe liver failure. Furthermore, do not just add glutamine to a regular diet. For every 1 gram of glutamine ingested, 1 gram of protein should be subtracted from the diet. For example, if you add 2 teaspoons (8 grams) of glutamine a day to your diet, you should remove 8 grams (approximately 1 ounce) of protein foods from your diet. Protein foods include meat, chicken, fish, eggs, cheese, milk, and yogurt.

You should also be careful about the foods to which you add your glutamine. Glutamine should be taken with cold or room-temperature foods or liquids. It should never be added to foods that are hot or highly acidic, such as vinegar. Heat destroys glutamine, as does acid. Glutamine added to a liquid should be taken within an hour of preparation or kept in a refrigerator or on ice. Fortunately, glutamine has no taste and readily dissolves in liquid, so powdered glutamine is easy to mix in water or flavored beverages or in foods such as cold puddings or applesauce.

RESTRICTIONS AND CONTRAINDICATIONS TO GLUTAMINE USE

Although glutamine is the most abundant amino acid in the human body and, as such, is natural to the body, care should still be exercised in its use. Some people should not take supplemental glutamine at all, while others should modify the dosage. For example, elderly people should probably not take the same dose of glutamine as younger people should take. As we age, the ability of our kidneys to do their job gradually declines, making it more and more difficult to process the toxic byproducts of glutamine metabolism.

Children have been given glutamine with no ill effects, but no child should be given any medication or supplement without supervision of a physician. In tests, glutamine has been given to children who have neither kidney nor liver disease in daily doses ranging from 0.25 to 0.5 gram for every 1 kilogram of body weight. More glutamine may be required during an acute illness, but the amount and its administration need to be integrated by the physican into the overall care program.

A pregnant woman with a condition that might be improved by glutamine may wonder if glutamine is safe for her unborn baby. Little, if any, research has been done on the use of glutamine during pregnancy. We do know that the glutamine level in amniotic fluid is high and that amniotic fluid is swallowed by the fetus throughout pregnancy.[3] Glutamine has also been given intravenously to very premature infants, who have displayed no untoward effects. In cows, the administration of glutamine has doubled milk production. If a pregnant woman had a condition that required glutamine therapy, most likely the supplement would be given, but her physician would have to evaluate her case on an individual basis.

Glutamine is being given to patients with cancer, but this is being done under closely supervised conditions, since glutamine is known to enhance the replication of cancer cells. Dr. Wiley Souba, now at Massachusetts General Hospital (see page 76), showed that animals with cancer that were given glutamine did not have larger tumors than did animals with cancer that were not given glutamine, but there was more tumor-cell division in the former, implying a more active tumor.[4] In spite of this, glutamine has been given to cancer patients receiving chemotherapy because glutamine has been shown to enhance the ability of the medications to kill cancerous growths.

However, glutamine use probably should be avoided by patients with chronic renal failure (kidney disease). In chronic renal failure, an excessive amount of amino acids or protein can be very detrimental to the kidneys and can cause further kidney damage. Considerable monitoring by a physician would be necessary if glutamine was given.

As discussed in Chapter 4, liver disease is another illness for which glutamine use may not be appropriate. Patients with severe liver disease should avoid glutamine supplementation.[5-7] Patients with severe cirrhosis of the liver, Reye's syndrome, or another metabolic disorder that can lead to an accumulation of ammonia in the blood are at increased risk for encephalopathy and coma. The basic problem in liver disease is an inability to properly clear the body of excess nitrogen, which is converted to ammonia and can ultimately cause brain swelling and brain-cell death. When the liver is *severely* damaged or when hepatic coma is imminent, glutamine is not effective and might cause further damage to the brain. In fact, many investigators now feel that because of the excess ammonia in the blood during these pathological states, the enzyme glutamine synthetase is stimulated to produce more glutamine in an effort to clear the blood of the ammonia. In these patients, the glutamine accumulates in the brain cells, draws water into the cells, causes the cells to swell, and eventually destroys them. This, of course, does cause permanent damage to the brain.

POSSIBLE SIDE EFFECTS OF GLUTAMINE

Glutamine has a side effect that might be annoying to some people. One major function of glutamine is to help transport water from the inside of the colon back into the body. This is very beneficial for someone with diarrhea, but it can be a problem for individuals who do not have diarrhea or who are prone to constipation. For example, someone with inflammatory bowel disease could greatly benefit from all of the positive effects of glutamine, but someone without diarrhea might find his or her stool becoming *too* firm. Eating a diet high in soluble fiber and drinking lots of water usually resolves the problem.

Soluble fiber prevents water from being reabsorbed into the body from the colon and rectum, so it counterbalances the effects that glutamine has on transporting water. Cereals such as Fiber One and All-Bran are high in soluble fiber, as are foods such as oat bran and apple pectin. Apple pectin can be found at health food stores. To normalize a stool made too firm by glutamine, begin by adding 1 gram of apple pectin to the diet per day along with plenty of water (at least 2 quarts of water per day for an adult). Increase the pectin up to 5 grams per day. In general, a quarter teaspoon of apple pectin equals 1 gram of soluble fiber. Many people mix the pectin into applesauce or pudding, or stir it into apple juice or another cold beverage.

Some people may wonder if glutamine is connected to monosodium glutamate (MSG) and the Chinese restaurant syndrome. The Chinese restaurant syndrome is caused by eating foods flavored with MSG. The symptoms include headache, chest pain, facial pressure, and tingling and burning sensations of the skin. Glutamine is related to MSG somewhat because of a similar biochemical structure, but the two compounds are quite different. Moreover, the symptoms of MSG use cannot be reproduced by taking glutamine. Conversely, the beneficial effects of glutamine cannot be achieved by taking MSG or glutamate (glutamic acid).

Although *glutamate* and *glutamic acid* sound similar to *glutamine*, the compounds are quite different. Glutamine is unique because it contains two nitrogen molecules. The amino acid glutamate, in contrast, contains only one nitrogen molecule and an extra acid group where the second nitrogen molecule is located in glutamine. It is the extra nitrogen that gives glutamine all of its special characteristics. Glutamate is not a substitute for glutamine in the body, and glutamate supplements do not provide any of the positive effects of glutamine.

Gluten is another protein whose name sounds quite similar to *glutamine*. In an indirect way, the two are related. Gluten, the protein in wheat, is made up of two protein fractions, gliadin and glutenin. In the gliaden portion, glutamine is one of the prominent amino acids. Many people have a negative reaction to wheat protein, or gluten. This negative reaction, however, is *not* related to the glutamine in the gluten but rather may be stimulated by the large structural protein of the substance.

A person, in fact, cannot be allergic to glutamine, which is the most abundant amino acid in the body. People are not allergic to individual amino acids, although they may experience negative effects from taking too large a dose of any amino acid. Glutamine is no exception, and it should be taken under the direction of a doctor. To date, no ill effects have been reported in people who, under the supervision of a doctor, have taken doses of 40 grams or less for weeks at a time.

Glutathione, another of the *glu* compounds, has a significant and close relationship to glutamine. As you may remember from page 43, glutathione is one of the most important compounds in the body because it scavenges and neutralizes the oxygen free radicals that cause so much destruction. Glutathione is made up of three amino acids—cysteine, glutamate, and glycine. When glutathione is taken orally, it is ineffective because it is absorbed directly into the enterocytes (the cells of the intestines) and broken down into its component parts. When it is transported to the liver, the liver, because of its complexities, uses a molecule of glutamate from glutamine to remanufacture glutathione, rather than using the original glutamate, which is absorbed. Studies have demonstrated that when the antioxidant glutathione is required, provision of glutamine along with cysteine will greatly enhance the synthesis of glutathione and facilitate more rapid recovery from illness.

It would be wonderful if we could go to our physicians and ask to have our glutamine and glutathione levels checked, since adequate levels are so important for good health. Unfortunately, readings of these levels are not very easy to obtain. As mentioned in Chapter 1, it took a long time for scientists to recognize the importance of glutamine partly because it is so hard to analyze in the laboratory. Even now, blood glutamine and glutathione levels are measured only as a research tool to help scientists understand more about glutamine. Furthermore, blood glutamine levels can be normal when total body levels are low. Muscle discharges glutamine into the bloodstream at times of stress to help maintain normal blood glutamine levels, but muscle glutamine levels become depleted.

Hopefully, however, measuring glutamine and glutathione levels will become routine in the future, much as checking blood pressure and knee-jerk reflexes is now. Hopefully, glutamine will be included in hospital intravenous solutions as a matter of course and will be prescribed by physicians for appropriate dysfunctions. And hopefully, this book will help to bring these things about, not just in the future, but in the near future.

Glossary

Acetaminophen. An analgesic, or pain medication; one brand is Tylenol.

Acquired immunodeficiency syndrome. A progressive, fatal disease caused by the human immunodeficiency virus (HIV).

Adenoids. Lymphatic tissues that resemble a gland and are found in the throat.

Adenosine triphosphate. A compound in living tissue that stores energy.

Adipose. Fat tissue in the body.

Adrenaline. A hormone secreted by the adrenal glands in times of stress.

AIDS. See *Acquired immunodeficiency syndrome.*

Amino acid. A molecule that contains at least one nitrogen group (a nitrogen atom linked to two hydrogen atoms) and one acid group connected in a specific manner; the amino acids are the building blocks of protein.

Ammonia. A soluble gas formed in the body as a breakdown waste product of protein.

Anabolism. The metabolic process in which food is converted into body tissue.

Analgesia. Relief of pain.

Antibody. A molecular compound formed by the immune system in response to a specific antigen; important in fighting infections.

Antidote. A compound that counteracts a poison.

Antigen. A substance that is perceived by the immune system as foreign, stimulating the immune system to form antibodies.

Antioxidant. A substance that prevents or delays oxygen free radicals from causing damage to cells.

Atherosclerosis. Hardening of arterial walls.

ATP. See *Adenosine triphosphate.*

Atrophy. Wasting of body tissue or cells.

Attenuate. To weaken, make less virulent.

Autoimmune. An immune response to your own body tissue because the immune system is unable to recognize self from nonself.

Bacterial translocation. In relation to the gastrointestinal tract, the movement of bacteria from inside the intestines through the bowel wall to the tissue surrounding the intestines.

Basal metabolism. The state of minimal metabolic activity, usually measured after a night's sleep, when the body is at physical and mental rest at normal body temperature.

B-cell. A type of lymphocyte important in the body's immune response; an immune cell that synthesizes humoral antibody.

Beta-carotene. A precursor to vitamin A; unlike vitamin A, it causes no toxicity when ingested in large doses.

Bile. A liquid produced by the liver and stored in the gallbladder that acts like a soap or detergent, taking large globules of fat and making them into smaller and smaller globules until they are small enough to be absorbed into the bloodstream through the small intestine.

Bone marrow. Blood-forming tissue located in the cavities of bones.

Bone-marrow transplantation. A procedure in which healthy bone marrow is removed from an individual and inserted into someone whose bone-marrow cells were killed by drugs or radiation; used to treat patients with blood cancer or to restore the bone marrow in patients treated for solid tumors.

Carbohydrate. Sugar or starch; a form of energy; an organic molecule composed of carbon, hydrogen, and oxygen that when broken down releases water, carbon dioxide, and energy.

Catabolism. The metabolic process in which body tissue is broken down into simpler compounds for energy.

Cell culture. The growth of cells outside the body in a dish filled with nutrients; the cells multiply but do not form recognizable tissue.

Chemical bond. A force or mechanism that links atoms into a molecule.

Chemotherapy. The use of drugs to treat a disease; usually used in cancer treatment.

Cholecystectomy. The removal of the gallbladder.

Cirrhosis. A liver disease in which normal tissue is replaced by fibrotic, nodular tissue.

Colitis. Inflammation of the large intestine.

Colon. A tubular structure that is the part of the large intestine extending from the small bowel to the rectum.

Conditionally essential amino acid. An amino acid that sometimes must be obtained externally and at other times is made by the body in sufficient quantities.

Cortisol. A hormone from the adrenal gland that is produced in the body during times of stress; called the stress hormone.

Crohn's disease. Inflammation of part of the bowel, usually the ileum, or lower part of the small intestine; can also occur in the colon.

Cysteine. The sulphur-containing amino acid essential in forming the antioxidant glutathione.

Cytidine. A molecule that is part of deoxyribonucleic acid (DNA).

Cytokine. A protein produced by immune cells that serves as a signal to warn the body that there is an infection or inflammation and that immune cells and certain nonimmune cells consequently need to alter their functions.

Deoxyribonucleic acid. The genetic material of life.

Detoxify. To eliminate the toxic nature of a compound.

Diverticular disease. A condition in the colon involving diverticula, or outpouchings, of the mucosa through the muscular layer of the colon wall; also known as diverticulosis.

Diverticulitis. Diverticular disease in which the diverticula are inflamed.

DNA. See *Deoxyribonucleic acid.*

Duodenum. The first of the three parts of the small intestine.

Edema. Tissue swelling.

EEG. See *Electroencephalography*.

Electroencephalography. The recording of the currents from nerve-cell conduction in the brain. The instrument used is called an electroencephalograph, and the record or chart is called an electroencephalogram.

Electron. A negatively charged particle that orbits around an atom; responsible for the bonding of different atoms.

Element. In chemical terms, a substance that cannot be decomposed any further by chemical means; examples are carbon, nitrogen, oxygen, and sulphur.

Emulsification. Dispersion of one liquid in another liquid by the formation of globules.

Encephalopathy. A condition usually associated with organ failure that produces loss of consciousness and may progress to deep coma.

Endocrine gland. A tissue that secretes a hormone into the bloodstream; the hormone then acts on distant tissue.

Endotoxin. A toxin derived from bacteria; causes increased temperature, decreased blood pressure, and leaky blood vessels.

Enterocyte. A cell that lines the small bowel.

Enzyme. A protein that acts on a substance to facilitate its conversion to another substance.

Epinephrine. See *Adrenaline*.

Essential amino acid. An amino acid that must be taken from a food source because it cannot be made by the body in sufficient quantity.

Estrogen. One of the two major feminizing hormones (the other is progesterone).

Exocrine gland. A tissue that secretes a substance through a duct to the surface of a tissue or organ or into a vessel; the substance then acts near its gland of origin.

Fat. A source of energy; an organic compound composed of hydrogen, carbohydrate, and oxygen.

Fatty liver. A liver permeated with fat; caused by excess alcohol consumption, dieting, or overeating.

5-fluorouracil. A cancer chemotherapeutic agent.

5-FU. See *5-fluorouracil.*

5-hydroxytryptamine. A vasoconstrictor found in many body tissues, including the central nervous tissue.

Free amino acid. An amino acid that is not bound to any other compound.

Free radical. An atom or group of atoms carrying an unpaired electron, making it very reactive and damaging to body tissue.

GABA. See *Gamma amino butyric acid.*

Gallbladder. The storage organ for bile; located under the liver.

Gamma amino butyric acid. An inhibitory neurotransmitter in the brain.

Gastritis. Inflammation of the lining of the stomach.

Gastrointestinal tract. The digestive system from the mouth to the anus.

GI tract. See *Gastrointestinal tract.*

Glucocorticoid. A hormone from the adrenal cortex that

stimulates the formation of glycogen and blood glucose.

Glucose. A simple sugar obtained from food or produced by carbohydrate digestion that is a source of energy for the body.

Glutamate. An amino acid similar to glutamine but lacking the extra nitrogen group.

Glutamic acid. See *Glutamate.*

Glutaminase. An enzyme that facilitates the splitting of glutamine into glutamate and ammonia.

Glutamine. A conditionally essential amino acid; contains two nitrogen groups, unlike the other amino acids, which contain one.

Glutamine synthetase. An enzyme that facilitates the production of glutamine from glutamate and ammonia.

Glutathione. A compound composed of glutamate, cysteine, and glycine that acts as an antioxidant in the body and prevents tissue damage.

Glycine. A nonessential amino acid.

Glycogen. The main carbohydrate stored in the body; found primarily in the liver.

Helper T-cell. A type of lymphocyte that has migrated to the thymus and matured there; helps B-cells and other T-cells to destroy foreign substances.

Hepatic artery. The artery bringing blood to the liver.

Hepatocyte. A cell of the liver.

HIV. See *Human immunodeficiency virus.*

Homocysteine. An amino acid that is important in the metabolism of cysteine and methionine but is not incorporated into proteins; elevated blood levels of homocysteine have been found to be associated with an increased incidence of heart disease.

Hormone. A substance in the body that is secreted into the bloodstream and acts at a site distant from its organ of origin to regulate the activity of an organ or group of cells.

Human immunodeficiency virus. A virus that prevents the immune system from mounting an immune response to fight infection; considered the causative factor of AIDS.

Hypermetabolism. Abnormally increased utilization of body materials; increased metabolism.

Immune response. The reaction of the body against a specific attack by microorganisms such as bacteria, fungi, or viruses.

Immune system. The system of responses that defends and protects the body from foreign material.

Immunoglobulin. Any of five specific proteins with antibody activity produced by lymphocytes and plasma cells in response to an antigen.

Immunosuppression. The inability of the body to mount an immune response.

Inflammatory bowel disease. Any of several illnesses marked by inflammation of the wall of the bowel, such as Crohn's disease and ulcerative colitis.

Intravenous. Within a vein; usually refers to infusion of a solution of water, dextrose (sugar), and salt to maintain adequate hydration in a patient who is unable to drink fluid because of illness or surgery.

IV. See *Intravenous.*

Kupffer's cell. A specialized immune cell in the liver that phagocytoses (eats) foreign material.

Leukocyte. The immune cell of the blood; the five types are lymphocytes, monocytes, neutrophils, basophils, and eosinophils.

Lumen. A cavity within a tube or organ.

Lumenal drive. Stimulation of the cells of the gastrointestinal tract to grow by the presence of digesting food in the gastrointestinal tract.

Lymph. A fluid that originates in many organs and tissues, and circulates throughout the body via the lymphatic vessels.

Lymph node. A small oval structure found in clusters in several areas of the body, including the mouth, neck, armpit, lower arm, and groin. It filters lymph, fights infection, and is the site of the formation of white blood cells and plasma cells.

Lymphocyte. The kind of white blood cell that initiates the immune response to a foreign substance, enabling all the other immune cells to move into action; the brains of the immune system.

Lymphoid. Pertaining to lymph, lymphatic tissue, or the lymphatic system.

Malnutrition. Abnormal nutrition; in developed nations, refers to overnutrition, which causes overweight, rather than undernutrition.

Metabolism. The sum of all the body processes that transform food into body matter (anabolism) or break down body matter for energy (catabolism).

Methotrexate. A chemotherapeutic agent used to fight cancer.

Microbe. A minute living organism such as a bacterium or fungus that may be capable of causing disease.

Mineral. An inorganic substance that is usually derived from the Earth's crust; many minerals are necessary for the proper functioning of the body and can be obtained from food.

Mortality. Death rate.

Mucomyst. A liquid or compound composed of acetyl cysteine, which acts as a donor of cysteine to help form glutathione in cases of acetaminophen overdose.

Mucosa. A membrane lining many body cavities that produces mucus.

Mucositis. Inflammation of a mucosa.

Myeloid. Pertaining to the bone marrow or spinal cord.

Necrosis. Death of tissue.

Neurotransmitter. A chemical substance produced by a nerve to act as a bridge to enable a nerve impulse to cross the space to the next nerve.

Nonessential amino acid. An amino acid that the body can make from other available nutrients.

Nonsteroidal anti-inflammatory drug. A medication that blunts the inflammatory response in the body.

NSAID. See *Nonsteroidal anti-inflammatory drug.*

Nutrient. A compound such as a protein, carbohydrate, fat, vitamin, or mineral that is necessary for the normal growth and functioning of the body.

Osteoporosis. A condition marked by a reduction in the quantity and quality of bone because of the loss of bone mineral and protein; bone thinning.

Oxygen free radical. A harmful substance produced when white blood cells use oxygen to make powerful chemicals to use against invading organisms; best known are the hydroxyl and peroxide molecules, which are similar to the lye that is used in household cleaning solutions, and hydrogen peroxide, which is used for bleaching and as a disinfectant.

Pancreas. An organ of the digestive system that serves both exocrine and endocrine functions.

Peptide. A small chain of amino acids.

Perfuse. To pour over or through, especially through the blood vessels of an organ.

Peyer's patches. The clusters of lymph nodes near the juncture of the small intestine and the colon.

Phagocyte. A cell that devours microorganisms or foreign material.

Phenobarbital. A medication used as an anticonvulsant, sedative, or hypnotic.

Placebo. An inert compound that appears identical to the active compound being tested in a research experiment; given to keep the test subject and/or researcher from prejudging the active compound being tested.

Plasma. The fluid portion of blood or lymph; plasma cells have a large nucleus, which under certain circumstances produces immunoglobulins.

Portal vein. A vein that drains blood from the digestive organs and spleen, and carries it to the liver.

Prednisone. A synthetic hormone used to treat severe inflammation.

Protein. A substance composed of carbohydrate, hydrogen, oxygen, and nitrogen; makes up the structural and functional tissue of the body.

RDA. See *Recommended dietary allowances.*

Recommended dietary allowances. The level of intake of essential nutrients that are judged by the Food and Nutrition Board to be adequate to meet the known nutrient needs of most healthy persons.

Red blood cell. The oxygen-carrying cell of the blood.

Reye's syndrome. A syndrome of encephalopathy and fatty degeneration of the liver, marked by rapid development of swelling of the brain and liver, and by disturbed consciousness and seizures; has been linked to chicken-pox, flu, stomach viruses, and Epstein-Barr virus.

Scavenger compound. A substance that affects chemical reactions in the body by readily combining with free radicals; glutathione is an example.

Scurvy. A disease caused by lack of vitamin C in the diet; characterized by soft, spongy, swollen, bleeding gums, as well as by easy bleeding and bruising of the skin.

Sepsis. The presence of pathologic organisms or their toxins in the blood or other tissues producing fever, chills, muscle aches, and sometimes a fall in blood pressure; infection.

Serotonin. See *5-hydroxytryptamine.*

Short bowel syndrome. A syndrome of malabsorption of nutrients and water that occurs after removal of all or part of the intestines.

Sinusoid. A terminal blood channel consisting of large, irregular vessels and found in many organs; often refers to the liver.

Spleen. A large glandlike organ of the lymphatic system, located in the upper left part of the abdomen.

Stem cell. An unspecialized cell that produces specialized cells, such as blood cells.

Stress response. The response of the body to external stimuli that includes the discharge of hormones primarily from the adrenal gland.

Striated muscle. Skeletal muscle that is attached to bone and generally crosses a joint; also called voluntary muscle.

Synthesize. To build up a compound by combining elements or other suitable starting materials.

T-cell. A type of lymphocyte that either passes through the thymus or is influenced by it; can assist or suppress the stimulation of antibody production in B-cells.

Thymus. A glandlike structure in the chest that is most active in early childhood and is related to the lymphatic system.

Tonsil. A mass of lymphoid tissue; often refers to the lymphoid tissue in the pharynx.

Total parenteral nutrition. A feeding solution of water plus all the nutrients considered necessary to sustain life, including fat, carbohydrate, amino acids, vitamins, and minerals.

Toxic oxidant. A chemical compound that carries a strong negative charge and is toxic to tissue with which it comes in contact.

TPN. See *Total parenteral nutrition.*

Transport. The movement of materials in biologic systems, particularly into and out of cells.

Tryptophan. An amino acid essential for optimal growth in children and protein equilibrium in adults.

Turnover. The growth of cells in a tissue to replace dead or dying cells.

Ulcer. An erosion of surface tissue caused by damage to the superficial cells.

Ulcerative colitis. An inflammatory bowel disease marked by ulcers of the colon.

Urea. The chief nitrogen breakdown product of protein metabolism, excreted in the urine.

Uridine. A compound that is part of DNA.

Villi. The many tiny vascular projections that cover the surface of the small intestine, and absorb and transport nutrients and fluids.

Vitamin. An organic substance found in many foods that is necessary for the normal metabolic functioning of the body.

White blood cell. See *Leukocyte.*

Notes

Chapter 1
Glutamine—Essential for Health

1. O'Dwyer, S.T., R.J. Smith, T.L. Hwang, D.W. Wilmore. "Maintenance of Small Bowel Mucosa With Glutamine-Enriched Parenteral Nutrition." *JPEN* 13 (1989):579–585.

2. Newsholme, E.A., M. Parry-Billings. "Properties of Glutamine Release From Muscle and Its Importance for the Immune System. *JPEN* 14 (suppl) (1990):63–67.

3. Okabe, S., K. Takeuchi, K. Honda, K. Takagi. "Effects of Acetylsalicylic Acid (ASA), ASA Plus L-Glutamine and L-Glutamine on Healing of Chronic Gastric Ulcer in the Rat." *Digestion* 14 (1976):85–88.

4. Rennie, M.J. "Muscle Protein Turnover and the Wasting Due to Injury and Disease." *Br Med Bull* 14 (1985): 257–264.

5. Rennie, M.J., et al. "Skeletal Muscle Glutamine Transport, Intramuscular Glutamine Concentration, and Muscle-Protein Turnover. *Metabolism* 38 (1989):47–51.

6. Souba, W.W., R.J. Smith, D.W. Wilmore. "Glutamine Metabolism by the Intestinal Tract." *JPEN* 9 (1985):608–617.

7. Ardawi, M.S.M., E.A. Newsholme. "Fuel Utilization in Colonocytes of the Rat." *Biochem J* 231 (1985):713–719.

8. Welbourne, T.C. "Interorgan Glutamine Flow in Metabolic Acidosis." *Am J Physiol* 250 (1986):E457–E463.

9. Ziegler, T.R., et al. "Clinical and Metabolic Efficacy of Glutamine-Supplemented Parenteral Nutrition After Bone Marrow Transplantation: A Double-Blind, Randomized, Controlled Trial." *Ann Intern Med* 116 (1992):821–828.

10. Szondy, Z., E.A. Newsholme. "The Effect of Time of Addition of Glutamine or Nucleosides on Proliferation of Rat Cervical Lymph-Node T-Lymphocytes After Stimulation by Concanavalin A." *Biochem J* 278 (1991):471–474.

11. Kabelic, T., et al. "Suppression of Human Immunodeficiency Virus Expression in Chronically Infected Monocyte Cells by Glutathione, Glutathione Ester, and N-Acetylcysteine." *Proc Natl Acad Sci USA* 88 (1991): 986–990.

12. Lacey, J.M., D.W. Wilmore. "Is Glutamine a Conditionally Essential Amino Acid?" *Nutr Rev* 48 (1990): 297–309.

13. Eagle, H., et al. "The Growth Response of Mammalian Cells in Tissue Culture to L-Glutamine and L-Glutamic Acid." *J Biol Chem* 218 (1956):607–616.

14. Windmueller, H.G. "Glutamine Utilization by the Small Intestine." *Adv Enzymol* 53 (1982):201–237.

15. Muhlbacher, F., et al. "Effects of Glucocorticoids on Glutamine Metabolism in Skeletal Muscle." *Am J Physiol* 247 (1984):E75–E83.

16. Johnson, D.J., et al. "Branched Chain Amino Acid Uptake and Muscle Free Amino Acid Concentrations Predict Postoperative Muscle Nitrogen Balance." *Ann Surg* 204 (1986):513–523.

17. Kapadia, C.R., et al. "Maintenance of Skeletal Muscle Intracellular Glutamine During Standard Surgical Trauma." *JPEN* 9 (1985):583–589.

18. Aulick, L.H., D.W. Wilmore. "Increased Peripheral Amino Acid Release Following Burn Injury." *Surgery* 85 (1979):560–565.

19. Stehle, P., et al. "Effects of Parenteral Glutamine Peptide Supplements on Muscle Glutamine and Nitrogen Balance After Major Surgery." *Lancet* i (1989):231–233.

Chapter 2
Preventing Muscle Breakdown

1. Bergström, J., et al. "Intracellular Free Amino Acid Concentration in Human Muscle Tissue." *J Appl Physiol* 36 (1974):693–701.

2. Stein, W.H., S. Moore. "The Free Amino Acids of Human Blood Plasma." *J Biol Chem* 211 (1954):915–926.

3. Roth, E., et al. "Metabolic Disorders in Severe Abdominal Sepsis, Glutamine Deficiency in Skeletal Muscle." *Clin Nutr* 1 (1982):25–41.

4. Rennie, M.J., et al. "Characteristics of Glutamine Carrier in Skeletal Muscle Have Important Consequences

for Nitrogen Loss in Injury, Infection, and Chronic Disease." *Lancet* i (1986):1008–1012.

5. Askanazi, J., et al. "Muscle and Plasma Amino Acids Following Injury. Influence of Intercurrent Infection." *Ann Surg* 192 (1980):78–85.

6. Caldwell, M.D. "Local Glutamine Metabolism in Wounds and Inflammation." *Metabolism* 38 (suppl) (1989): 34–39.

7. Parry-Billings, M., J. Evans, P.C. Calder, E.A. Newsholme. "Does Glutamine Contribute to Immunosuppression After Major Burns?" *Lancet* 336 (1990):523–525.

8. Newsholme, E.A., M. Parry-Billings. "Properties of Glutamine Release From Muscle and Its Importance for the Immune System." *JPEN* 14 (suppl 4) (1990):63–67.

9. Ardawi, M.S.M., E.A. Newsholme. "Glutamine, the Immune System, and the Intestine." *J Lab Clin Med* 115 (1990):654–655.

10. Souba, W.W., R.J. Smith, D.W. Wilmore. "Glutamine Metabolism by the Intestinal Tract." *JPEN* 9 (1985):608–617.

11. Windmueller, H.G. "Glutamine Utilization by the Small Intestine." *Adv Enzymol* 53 (1982):201–237.

12. Muhlbacher, F., et al. "Effects of Glucocorticoids on Glutamine Metabolism in Skeletal Muscle." *Am J Physiol* 247 (1984):E75–E83.

13. Kapadia, C.R., F. Muhlbacher, R.J. Smith, D.W. Wilmore. "Alterations in Glutamine Metabolism in Response to Cooperative Stress and Food Deprivation." *Surg Forum* 33 (1982):19–21.

14. Brooks, D.C., P.Q. Bessey, P.R. Black, T.T. Aoki, D.W. Wilmore. "Insulin Stimulates Branched Chain Amino Acid Uptake and Diminishes Nitrogen Flux From Skeletal Muscle of Injured Patients." *J Surg Res* 40 (1986): 395–405.

15. Marliss, E.B., T.T. Aoki, T. Pozefsky, A.S. Most, G.F. Cahill. "Muscle and Splanchnic Glutamine and Glutamate Metabolism in Postabsorptive and Starved Men." *J Clin Invest* 50 (1971):814–817.

16. Dudrick, S.J., D.W. Wilmore, H.M. Vars, J.E. Rhoads. "Long-Term Total Parenteral Nutrition With Growth, Development, and Positive Nitrogen Balance." *Surgery* 64 (1968):134–142.

17. Street, S.J., A.H. Beddoe, G.L. Hill. "Aggressive Nutritional Support Does Not Prevent Protein Loss Despite Fat Gain in Septic Intensive Care Patients." *J Trauma* 27 (1987): 262–266.

18. Ziegler, T.R., et al. "Clinical and Metabolic Efficacy of Glutamine-Supplemented Parenteral Nutrition After Bone Marrow Transplantation." *Ann Int Med* 116 (1992): 821–828.

19. Stehle, P., et al. "Effects of Parenteral Glutamine Peptide Supplements on Muscle Glutamine Loss and Nitrogen Balance After Major Surgery." *Lancet* i (1987): 231–233.

20. Hammarqvist, F., et al. "Addition of Glutamine to Total Parenteral Nutrition After Elective Abdominal Surgery Spares Free Glutamine in Muscle, Counteracts the Fall in Muscle Protein Synthesis, and Improves Nitrogen Balance." *Ann Surgery* 209 (1989):455–461.

21. MacLennan, P.A., R.A. Brown, M.J. Rennie. "A Posi-

tive Relationship Between Protein Synthetic Rate and Intracellular Concentration in Perfused Rat Skeletal Muscle." *FEBS Lett* 215 (1987):187–191.

22. MacLennan, P.A., et al. "Inhibition of Protein Breakdown by Glutamine in Perfused Rat Skeletal Muscle." *FEBS Lett* 237 (1988):133–136.

23. Jepson, M.M., et al. "Relationship Between Glutamine Concentration and Protein Synthesis in Rat Skeletal Muscle." *Am J Physiol* 355 (1988):E166–E172.

24. Welbourne, T.C. "Enteral Glutamine Spares Endogenous Glutamine in Chronic Acidosis." *JPEN* 17 (1993): 23S.

Chapter 3
Healing the Stomach and Intestines

1. Helton, W.S., R.J. Smith, J. Rounds, D.W. Wilmore. "Glutamine Prevents Pancreatic Atrophy and Fatty Liver During Elemental Feeding." *J Surg Res* 48 (1990):297–303.

2. Wilmore, D.W. "The Gut: A Central Organ After Surgical Stress." *Surgery* 104 (1988):917–923.

3. Windmueller, H.G., A.E. Spaeth. "Identification of Ketone Bodies and Glutamine as the Major Respiratory Fuels In Vivo for Postabsorptive Rat Small Intestine." *J Biol Chem* 253 (1978):69–76.

4. Hartmann, F., M. Plaath. "Intestinal Glutamine Metabolism." *Metabolism* 38 (suppl 1) (1989):18–24.

5. Asby, A.A., M.S.M. Ardawi. "Glucose, Glutamine, and Ketone Body Metabolism in Human Enterocytes." *Metabolism* 37 (1988):602–609.

6. Souba, W.W., R.J. Smith, D.W. Wilmore. "Glutamine Metabolism by the Intestinal Tract." *JPEN* 9 (1985):608–617.

7. Deitch, E.A., et al. "The Gut as a Portal Entry for Bacteremia." *Ann Surg* 205 (1987):681–692.

8. Mochizuki, H. "Mechanism of Prevention of Post Burn Hypermetabolism and Catabolism by Early Enteral Feeding." *Ann Surg* 200 (1984):297–310.

9. O'Dwyer, S.T., R.J. Smith, T. Hwang, D.W. Wilmore. "Maintenance of Small Bowel Mucosa With Glutamine-Enriched Parenteral Nutrition." *JPEN* 13 (1989): 579–585.

10. O'Dwyer, S.T., et al.. "5-Fluorouracil Toxicity on Small Intestinal Mucosa But Not White Blood Cells Is Decreased by Glutamine." *Clin Res* 35 (1987):369A.

11. Jacobs, D.O., et al. "Disparate Effects of 5-Fluorouracil on the Ileum and Colon of Enterally Fed Rats With Protection by Dietary Glutamine." *Surg Forum* 38 (1987): 45–47.

12. Fox, A.D., et al. "Effects of Glutamine-Supplemented Enteral Diet on Methotrexate-Induced Enterocolitis." *JPEN* 12 (1988):325–331.

13. Fox, A.D., et al. "Reduction of the Severity of Enterocolitis by Glutamine-Supplemented Enteral Diets." *Surg Forum* 38 (1987):43–44.

14. Klimberg, V.S., et al. "Prophylactic Glutamine Protects the Intestinal Mucosa From Radiation Injury." *Cancer* 66 (1990):62–68.

15. Ziegler, T.R., et al. "Clinical and Metabolic Efficacy of Glutamine-Supplemented Parenteral Nutrition After

Bone Marrow Transplantation." *Ann Intern Med* 116 (1992):821–828.

16. Okabe, S., K. Takeuchi, K. Honda, K. Takagi. "Effects of Acetylsalicylic Acid (ASA), ASA Plus L-Glutamine and L-Glutamine on Healing of Chronic Gastric Ulcer in the Rat." *Digestion* 14 (1976):85–88.

17. Rhoads, J.M., E.O. Keku, J. Quinn, J. Wooseley, J.G. Lecce. "L-Glutamine Stimulates Jejunal Sodium and Chloride Absorption in Pig Rotavirus Enteritis." *Gastroenterology* 100 (3) (1991):683–691.

18. Ardawi, M.S.M., E.A. Newsholme. "Fuel Utilization in Colonocytes of the Rat." *Biochem J* 231 (1985):713–719.

19. Belli, D.C., et al. "Chronic Intermittent Elemental Diet Improves Growth Failure in Children With Crohn's Disease." *Gastroenterology* 94 (1988):603–610.

20. Yoshimura, K., et al. "Effect of Enteral Glutamine Administration on Experimental Inflammatory Bowel Disease." Abstract. *JPEN* 17 (1993):235.

21. Weir, C.D., et al. "The Effect of Glutamine Supplemented Elemental Diet on Disease Activity in a Chronic Colitis Model." Abstract. *JPEN* 17 (1993):345.

22. Byrne, T.A., T.B. Morrissey, T.R. Ziegler, C. Gatzen, L.S. Young, D.W. Wilmore. "Growth Hormone, Glutamine and Fiber Enhance Adaptation of Remnant Bowel Following Massive Intestinal Resection." *Surg Forum* 43 (1992):151–153.

23. Wischmeyer, P., R.L. Grotz, J.H. Pemberton, S.F. Phillips. "Treatment of Pouchitis After Ileo-Anal Anastomosis With Glutamine and Butyric Acid." Abstract. *Gastroenterology* (April 1992):A947.

24. Wilmore, D.W. Personal communication, December 1992.

25. Wilmore, D.W. "The Effect of Glutamine on the Gastrointestinal Tract." *Rivista Ital di Nutriz Parent ed Enter* 10 (1992):1–6.

Chapter 4
Supporting the Liver

1. Häussinger, D. "Glutamine Metabolism in the Liver: Overview and Current Concepts." *Metabolism* 38 (suppl 1) (1989):14–17.

2. Meister, A. "Glutathione Metabolism and Its Selective Modification." *J Biol Chem* 263 (1988):17205–17208.

3. Taniguchi, N., T. Higashi, Y. Sakamoto, A. Meister. *Glutathione Centennial: Molecular Properties and Clinical Applications.* (New York: Academic Press, 1989).

4. Freeman, B.A., J.D. Crapo. "Biology of Disease: Free Radicals and Tissue Injury." *Lab Invest* 47 (1982): 412–426.

5. Weiss, S.J. "Tissue Destruction by Neutrophils." *N Engl J Med* 320 (1989):365–376.

6. Southorn, P.A., G. Powis. "Free Radicals in Medicine. II. Involvement in Human Disease." *Mayo Clin Proc* 63 (1988):390–408.

7. Bulkley, G.B. "The Role of Oxygen Free Radicals in Human Disease Processes." *Surgery* 94 (1983):407–411.

8. Lyons, M.J., J.F. Gibson, D.J.E. Ingram. "Free Radicals Produced in Cigarette Smoke." *Nature* 181 (1958):1003–1004.

9. Smilkstein, M.J., G.L. Knapp, K.W. Kulig, B.H. Rumack. "Efficiency of Oral N-Acetylcysteine in the Treatment of Acetominophen Overdose. Analysis of the National Multicenter Study (1976 to 1985)." *N Engl J Med* 319 (1988): 1557–1562.

10. Hong, R.W., W.S. Helton, J.D. Rounds, D.W. Wilmore. "Glutamine-Supplemented TPN Preserves Hepatic Glutathione and Improves Survival Following Chemotherapy." *Surg Forum* 41 (1990):9–11.

11. Hong, R.W., J.D. Rounds, W.S. Helton, M.K. Robinson, D.W. Wilmore. "Glutamine Preserves Liver Glutathione After Lethal Hepatic Injury." *Ann Surg* 215 (1992):114–119.

12. Hong, R.W., M.K. Robinson, J.D. Rounds, D.W. Wilmore. "Glutamine Protects the Liver Following Corynebacterium Parvum/Endotoxin-Induced Hepatic Necrosis." *Surg Forum* 42 (1991):1–3.

13. Robinson, M.K., et al. "Glutathione Depletion Enhances Bacterial Translocation and Alters Immunologic Status." *Surg Forum* 42 (1991):65–67.

14. Kalebic, T., et al. "Suppression of Human Immunodeficiency Virus Expression in Chronically Infected Monocytic Cells by Glutathione, Glutathione Ester, and N-Acetylcysteine." *Proc Natl Acad Sci* 88 (1991): 986–990.

15. Robinson, M.K., R.W. Hong, D.W. Wilmore. "Glutathione Deficiency and HIV Infection." Letter to the Editor. *Lancet* 339 (1992):1603–1604.

16. Buhl, R., et al. "Systemic Glutathione Deficiency in Symptom-Free HIV-Seropositive Individuals." *Lancet* ii (8675) (1989):1294–1298.

17. Stall, F.J.T., et al. "Glutathione Deficiency and Human Immunodeficiency Virus Infection." *Lancet* 339 (1992): 909–912.

18. Welbourne, T.C., A.B. King, K. Hunter. "Enteral Glutamine Supports Hepatic Glutathione Efflux During Inflammation." *J Nut Biochem* 4 (1993):236–242.

19. Helton, W.S., R.J. Smith, J. Rounds, D.W. Wilmore. "Glutamine Prevents Pancreatic Atrophy and Fatty Liver During Elemental Feeding." *J Surg Res* 48 (1990): 297–303.

20. Li, J., L.H. Stahlgren. "Glutamine Prevents the Biliary Lithgenic Effect of Total Parenteral Nutrition in Rats." *JPEN* 17 (1993):28S.

Chapter 5
Strengthening the Immune System

1. Eagle, H., V.I. Oyama, M. Levy, C.L. Horton, R. Fleischman. "The Growth Response of Mammalian Cells in Tissue Culture to L-Glutamine and L-Glutamic Acid." *J Biol Chem* 218 (1956):607–616.

2. Eagle, H. "Nutrition Needs of Mammalian Cells in Tissue Culture." *Science* 122 (1955):501–504.

3. Newsholme, E.A., B. Crabtree, M.S.M. Ardawi. "Glutamine Metabolism in Lymphocytes: Its Biochemical, Physiological and Clinical Importance." *Q J Exper Physiol* 70 (1985):473–489.

4. Alverdy, J.C. "Effects of Glutamine-Supplemented Diets on Immunology of the Gut." *JPEN* 14 (1990):109S–113S.

5. Austgen, T.R., R. Chakrabarti, M.K. Chen, W.W. Souba.

"Adaptive Regulation in Skeletal Muscle Glutamine Metabolism in Endotoxin-Treated Rats." *J Trauma* 32 (1992):600.

6. Inoue, Y., J.P. Grant, P.J. Snyder. "Effect of Glutamine-Supplemented Intravenous Nutrition on Survival After Escherichia Coli-Induced Peritonitis." *JPEN* 17 (1993):41–46.

7. Baskerville, A., P. Hambleton, J.E. Benbough. "Pathological Features of Glutaminase Toxicity." *Br J Exp Pathol* 61 (1980):132–138.

8. Jacobs, D.O., A. Evans, S.T. O'Dwyer, R.J. Smith, D.W. Wilmore. "Disparate Effects of 5-Fluorouracil on the Ileum and Colon of Enterally Fed Rats With Protection by Dietary Glutamine." *Surg Forum* 38 (1987):45–47.

9. Parry-Billings, M., J. Evans, P.C. Calder, E.A. Newsholme. "Does Glutamine Contribute to Immunosuppression After Major Burns?" *Lancet* 336 (1990): 523–525.

10. Gottschlich, M., C. Powers, J. Khoury, G. Narden. "Incidence and Effects of Glutamine Depletion in Burn Patients." Abstract. *JPEN* 17 (1993): 235.

11. Hong, R.W., M.K. Robinson, J.D. Rounds, D.W. Wilmore. "Glutamine Protects the Liver Following Corynebacterium Parvum/Endotoxin-Induced Hepatic Necrosis." *Surg Forum* 42 (1991):1–3.

12. Ziegler, T.R., et al. "Clinical and Metabolic Efficacy of Glutamine-Supplemented Parenteral Nutrition After Bone Marrow Transplantation." *Ann Intern Med* 116 (1992):821–828.

13. Schloerb, P.R., M. Amare. "Total Parenteral Nutrition With Glutamine in Bone Marrow Transplantation and Other Clinical Applications (A Randomized, Double-Blind Study)." *JPEN* 17 (1993):404–413.

14. Young, L.C., C. Gatzen, K. Wilmore, D.W. Wilmore. "Glutamine (Gln) Supplementation Fails to Increase Plasma Gln Levels in Asymptomatic HIV+ Individuals." *JADA* 92 (suppl) (1992):A-88.

15. Robinson, M.K., R.W. Hong, D.W. Wilmore. "Glutathione Deficiency and HIV Infection." Letter to the Editor. *Lancet* 339 (1992):1603–1604.

16. Parry-Billings, M., et al. "Plasma Amino Acid Concentrations in the Overtraining Syndrome: Possible Effects on the Immune System." *Med Sci Sports Exerc* 24 (1992): 1353–1358.

17. Welbourne, T.C. "Enteral Glutamine Spares Endogenous Glutamine in Chronic Acidosis." *JPEN* 17 (1993):23S.

Chapter 6
Helping Against Depression, Anger, and Fatigue

1. Takahashi, K. "Studies on the Free Amino Acid Patterns in Cerebrospinal Fluid of Children." *Acta Paediatr Jpn Oversea Ed (Tokyo)* 20 (1978):11–23.

2. Shank, R.P., G.L. Campbell. "Metabolic Precursors of Glutamate and GABA." In Hortz, L., et al. *Glutamine, Glutamate and GABA in the Central Nervous System.* (New York: Liss, 1983), 355–369.

3. Rogers, L.L., R.B. Pelton, R.J. Williams. "Amino Acid Supplementation and Voluntary Alcohol Consumption by Rats." *J Biol Chem* 220 (1956):321–323.

4. Rogers, L.L., R.B. Pelton. "Glutamine in the Treatment of Alcoholism." *Quart St* 18 (1957):581–587.

5. Manna, V., N. Martucci. "Effects of Short-Term Administration of Cytidine, Uridine and Levoglutamine, Alone or in Combination, on the Cerebral Electrical Activity of Patients With Chronic Cerebrovascular Disease." *Int J Clin Pharm Res* 8 (1988): 199–210.

6. Rogers, L.L., R.B. Pelton. "Effects of Glutamine on IQ Scores of Mentally Deficient Children." *Tex Rep Biol Med* 15 (1957):84–90.

7. Cocchi, R. "Antidepressive Properties of L-Glutamine." *Acta Psychiatr Belg* 76 (1976):658–666.

8. Ziegler, T.R., et al. "Safety and Metabolic Effects of L-Glutamine Administration in Humans." *JPEN* 14 (suppl) (1990):137S–146S.

9. Young, L.S., M. Scheltinga, R. Bye, D.W. Wilmore. "Can Tests of Patient Well-Being Be Used to Evaluate Nutritional Efficacy? An Affirmative Answer." *JPEN* 16 (1992):20S.

10. Young, L.S., et al. "Patients Receiving Glutamine-Supplemented Intravenous Feedings Report an Improvement in Mood." *JPEN* 17 (1993):422–427.

11. Jain, P., N.K. Khanna. "Evaluation of Anti-Inflammatory and Analgesic Properties of L-Glutamine." *Agen Act* 11 (1981):243–249.

12. Wilmore, D.W. Personal communication, May 1992.

Chapter 7
Fighting Cancer

1. Austgen, T.R., P.S. Dudrick, H. Sitren, K.I. Bland, E.

Copeland, W.W. Souba. "The Effects of Glutamine-Enriched Total Parenteral Nutrition on Tumor Growth and Host Tissues." *Ann Surg* 215 (2) (1992):107–113.

2. Souba, W.W. "Glutamine and Cancer." *Ann Surg* 218 (1993):715–728.

3. Rouse, K., E.C. Nwokedi, J. Woodliff, J. Epstein, V.S. Klimberg. "Glutamine Enhances Selectivity of Chemotherapy Through Changes in Glutathione Metabolism." *JPEN* 17 (1993):28S.

4. Klimberg, V.S., et al. "Glutamine Facilitates Chemotherapy While Reducing Toxicity." *JPEN* 16 (suppl 1) (1992):83S–87S.

5. Klimberg, V.S., et al. "Effect of Supplemental Dietary Glutamine on Methotrexate Concentrations in Tumors." *Arch Surg* 127 (1992):1317–1320.

6. Klimberg, V.S. Personal communication, October 1993.

Chapter 8
Making Aging Easier

1. Miller, A.J., J.A. Brunell, J.P. Carlos, L.J. Brown, H. Loe. "Oral Health of United States Adults. National Findings." NIH Publication No. 87-2868, 1987.

2. Breustedt, A. "Age-Induced Changes in the Oral Mucosa and Their Therapeutic Consequences." *Int Dent J* 33 (1983):272–280.

3. Kulber, D.A., S. Hartunian, D. Schiller, L. Morgenstern. "The Current Spectrum of Peptic Ulcer Disease in the Older Age Groups." *Ann Surg* 56 (1990):737–741.

4. Miller, D.K. "Acute Upper Gastrointestinal Bleeding in Elderly Persons." *J Am Geriatr Soc* 39 (1991):409–422.

5. Cheskin, L.J., M. Bohlman, M.M. Schuster. "Diverticular Disease in the Elderly." *Gastroenterol Clin North Am* 19 (1990):391–403.

6. Wilmore, L.H. "The Aging of Bone and Muscle." *Clin Sports Med* 10 (1991):231–244.

7. Borkan, G.A., D.E. Hullts, S.G. Gerzof, A.H. Robbins, C.K. Silbert. "Age Changes in Body Composition Revealed by Computed Tomography." *J Gerontol* 38 (1983): 673–677.

8. Schneider, E.L., et al. "Recommended Dietary Allowances and the Health of the Elderly." *N Engl J Med* 314 (1986):157–160.

9. Hartz, S.C., R.M. Russell, I.H. Rosenberg, eds. *Nutrition in the Elderly. The Boston Nutritional Status Survey.* (London: Smith, Gordon, 1992).

10. Holick, M.F. "Vitamin D Requirements for the Elderly." *Clin Nutr* 5 (1986):121–129.

11. Krall, E.A., et al. "Effects of Vitamin D Intake on Seasonal Variations in Parathyroid Hormone Secretion in Postmenopausal Women." *N Engl J Med* 321 (1989): 1777–1783.

12. Russell, R.M. "Atrophic Gastritis." In Hartz, S.C., R.M. Russell, I.H. Rosenberg, eds. *Nutrition in the Elderly. The Boston Nutritional Status Survey.* (London: Smith, Gordon, 1992), 189–193.

13. Clarke, R., et al. "Hyperhomocysteinemia: An Independent Risk Factor for Vascular Disease." *N Engl J Med* 324 (1991):1149–1155.

14. Ribaya-Mercado, J.D., et al. "Vitamin B6 Requirement of Elderly Men and Women." *J Nutr* 121 (1991):1062–1074.

15. Meydani, S.N., et al. "Vitamin B6 Deficiency Impairs Interleukin-2 Production and Lymphocyte Proliferation in Elderly Adults." *Am J Clin Nutr* 53 (1991):1275–1280.

16. Bell, I.R., et al. "Plasma Homocysteine in Vascular Disease and in Nonvascular Dementia of Depressed Elderly People." *Acta Psych Scan* 85 (1992):386–390.

17. Chandra, R.K. "Effect of Vitamin and Trace-Element Supplementation on Immune Responses and Infection in Elderly Subjects." *Lancet* 340 (1992):1124–1127.

18. Meydani, S.N., et al. "Vitamin E Supplementation Enhances Cell-Mediated Immunity in Healthy Elderly." *Am J Clin Nutr* 52 (1990):557–563.

19. Stampfer, M.J., C.H. Hennekens, J.E. Manson, G.A. Colditz, B. Rosner. "Vitamin E Consumption and the Risk of Coronary Heart Disease in Women." *N Engl J Med* 328 (1993):1444–1449.

20. Rimm, E.B., et al. "Vitamin E Consumption and the Risk of Coronary Heart Disease in Men." *N Engl J Med* 328 (1993):1450–1456.

21. Murata, A. "Smoking and Vitamin C." *World Rev Nutr Biet* 64 (1991):31–57.

22. Schectman, G., J.C. Byrd, R. Hoffman. "Ascorbic Acid Requirements for Smokers: Analysis of a Population Survey." *Am J Clin Nutr* 53 (1991):1466–1470.

23. Varma, S.D. "Scientific Basis for Medical Therapy of Cataracts by Antioxidants." *Am J Clin Nutr* 53 (1991): 335S–345S.

24. Enstrom, J.F. "Vitamin C Intake and Mortality Among

a Sample of the United States Population." *Epidemiology* 3 (1992):194–202.

25. Chen, M.F., H.W. Boyce, J.M. Hsu. "Effect of Ascorbic Acid in Plasma Alcohol Clearance." *J Am Coll Nutr* 9 (1990):185–189.

26. Basu, J., et al. "Plasma Ascorbic Acid and Beta Carotene Levels in Women Evaluated for HPV Infection, Smoking, and Cervix Dysplasia." *Cancer Detect Prev* 15 (1991): 165–170.

27. Basu, J., et al. "Smoking and the Antioxidant Ascorbic Acid: Plasma, Leukocyte, and Cervicovaginal Cell Concentrations in Normal Healthy Women." *Am J Obstet Gynecol* 163 (1990):1948–1952.

28. Garewal, H.S. "Potential Role of Beta-Carotene in Prevention of Oral Cancer." *Am J Clin Nutr* 53 (1991): 294S–297S.

29. Ziegler, R.G. "Vegetables, Fruits and Carotenoids and the Risk of Cancer." *Am J Clin Nutr* 53 (1991):251S–259S.

30. Dorgan, J.F., A. Schatzkin. "Antioxidant Micronutrients in Cancer Prevention." *Hematol Oncol Clin North Am* 5 (1991):43–68.

31. Singh, V.N., S.K. Gaby. "Premalignant Lesions: Role of Antioxidant Vitamins and Beta-Carotene in Risk Reduction and Prevention of Malignant Transformation." *Am J Clin Nutr* 53 (1991):386S.

Chapter 9
Adding Glutamine to Your Life

1. Lowe, D.K., et al. "Glutamine-Enriched Parenteral Nu-

trition Is Safe in Normal Humans." *Surg Forum* 40 (1989):9–11.

2. Ziegler, T.R., et al. "Safety and Metabolic Effects of L-Glutamine Administration in Humans." *JPEN* 14 (1990):137S–146S.

3. Emery, A.E.H., D. Burt, M.M. Nelson, J.B. Scrimgeour. "Antenatal Diagnosis and Amino Acid Composition of Amniotic Fluid." *Lancet* i (1970):1307–1308.

4. Austgen, T.R., et al. "The Effects of Glutamine-Enriched Total Parenteral Nutrition on Tumor Growth and Host Tissues." *Ann Surg* 215 (1992):107–113.

5. Fraser, C.L., A.I. Arieff. "Hepatic Encephalopathy." *N Engl J Med* 313 (1985):865–873.

6. Takahashi, H., R. Koehler, S. Brusilow, R. Traystmen. "Inhibition of Brain Glutamine Accumulation Prevents Cerebral Edema in Hyperammonemic Rats." *Am J Physiol* 261 (1991):H825–H829.

7. Hawkins, R.A., J. Jessy. "Hyperammonemia Does Not Impair Brain Function in the Absence of Net Glutamine Synthesis." *Biochem J* 277 (1991):697–703.

About the Authors

Dr. Judy Shabert is an obstetrician/gynecologist with a medical degree from the University of Hawaii. She is also a registered dietician with a degree from North Dakota State University, and she holds a Master of Public Health degree from Harvard University. She completed her medical training at the New England Medical Center and her dietetic internship at Peter Bent Brigham Hospital, both in Boston. She is a consultant in health-related fields. Her husband, Doug Wilmore, MD, the Frank Sawyer professor of surgery at Harvard Medical School and one of the pioneers in parenteral nutrition at the University of Pennsylvania, is on the staff at Brigham and Women's Hospital in Boston, where he conducts glutamine research. This allows Judy to follow the progress of glutamine research very closely and gives her ready access to the scientific data and medical reports. Both the author and her husband are members of the American Dietetic Association. They reside in Brookline, Massachusetts.

Nancy Ehrlich has been a full-time writer for *Scientific American* for more than ten years. She is currently a consulting editor in *Scientific American*'s medical publishing division. She resides in New York City.

Index

Healthy Habits

are easy to come by—

IF YOU KNOW WHERE TO LOOK!

Get the latest information on:

- **better health • diet & weight loss**
- **the latest nutritional supplements**
- **herbal healing • homeopathy and more**

COMPLETE AND RETURN THIS CARD RIGHT AWAY!

Where did you purchase this book?

❑ bookstore ❑ health food store ❑ pharmacy
❑ supermarket ❑ other (please specify)_____

Name_____

Street Address_____

City_____State_____Zip_____

HEALTH BOOK CATALOG

ACHIEVING HEALTH

AVERY PUBLISHING GROUP

1998

RECEIVE A FREE COPY OF AVERY'S HEALTH CATALOG

GIVE ONE TO A FRIEND ...

Healthy Habits

are easy to come by—

IF YOU KNOW WHERE TO LOOK!

Get the latest information on:

- **better health • diet & weight loss**
- **the latest nutritional supplements**
- **herbal healing • homeopathy and more**

COMPLETE AND RETURN THIS CARD RIGHT AWAY!

Where did you purchase this book?

❑ bookstore ❑ health food store ❑ pharmacy
❑ supermarket ❑ other (please specify)_____

Name_____

Street Address_____

City_____State_____Zip_____

HEALTH BOOK CATALOG

ACHIEVING HEALTH

AVERY PUBLISHING GROUP

1998

RECEIVE A FREE COPY OF AVERY'S HEALTH CATALOG

Avery Publishing Group

120 Old Broadway
Garden City Park, NY 11040

|..ll...llll...l..lll...l.l.ll....ll.l.l...lll

||| ||

Avery Publishing Group

120 Old Broadway
Garden City Park, NY 11040

|..ll...llll...l..lll...l.l.ll....ll.l.l...lll